IEE Control Engineering Series
Series Editors: G.A. Mongomerie
Prof. H. Nicholson

Design of modern control systems

Editors: D.J. Bell,
P.A. Cook, N. Munro

Previous volumes in this series:

Volume 1	Multivariable control theory J.M. Layton
Volume 2	Lift traffic analysis, design and control G.C. Barney and S.M. dos Santos
Volume 3	Transducers in digital systems G.A. Woolvet
Volume 4	Supervisory remote control systems R.E. Young
Volume 5	Structure of interconnected systems H. Nicholson
Volume 6	Power system control M.J.H. Sterling
Volume 7	Feedback and multivariable systems D.H. Owens
Volume 8	A history of control engineering, 1800-1930 S. Bennett
Volume 9	Modern approaches to control system design N. Munro (Editor)
Volume 10	Control of time delay systems J.E. Marshall
Volume 11	Biological systems, modelling and control D.A. Linkens
Volume 12	Modelling of dynamical systems-1 H. Nicholson (Editor)
Volume 13	Modelling of dynamical systems-2 H. Nicholson (Editor)
Volume 14	Optimal relay and saturating control system synthesis E.P. Ryan
Volume 15	Self-tuning and adaptive control theory and application C.J. Harris and S.A. Billings (Editors)
Volume 16	Systems modelling and optimisation P. Nash
Volume 17	Control in hazardous environments R. E. Young

Design of modern control systems

Editors: D. J. Bell,
P. A. Cook, N. Munro

Peter Peregrinus Ltd.
on behalf of the
Institution of Electrical Engineers

Peter Peregrinus Ltd. on behalf of the Institution
of Electrical Engineers

Published by: Peter Peregrinus Ltd.,Stevenage and
New York

British Library Cataloguing in Publication Data.

Bell, D.J.
Design of modern control systems.
(IEE control engineering series; v.18)
1. Control theory
I. Title II. Cook, P.A. III. Munro, N.

ISBN 0-906048174-5

Printed in England by Short Run Press Ltd., Exeter.

Contents

Preface

Over the last seven years the Science and Engineering Research Council (previously the Science Research Council) have been instrumental in the organization and financing of a series of vacation schools. These have been aimed primarily at recipients of SERC postgraduate awards but some non-SERC students, together with a few workers from industry (in the UK and elsewhere) have also taken part. Two of these vacation schools have been on the subject of analysis and design of control systems and have been held at the Control Systems Centre in the University of Manchester Institute of Science and Technology (UMIST). The chapters of this present book form the lectures to be given at the third UMIST vacation school during the week 29 March — 3 April, 1982. The title of the vacation school is the same as that for this book.

It is inevitable and indeed desirable that a subject such as engineering design of multivariable systems will develop over a period of six years. This becomes apparent if one compares the contents of the 1976 UMIST vacation school as given in "Modern approaches to control system design" (N. Munro, Ed., IEE Control Engineering Series 9, Peter Peregrinus, 1979) with the contents of the volume before you. Some of the 1976 topics no longer appear in this 1982 version, not because they are now obsolete or research is no longer taking place in those areas but because we have felt it right and proper to make way for either new subjects (such as large scale systems) or interesting new applications (e.g. the nuclear boiler in Chapter 6). Even in the case where 1976 topics are retained we have arranged for some different aspects to be presented in this new publication. It will be noticed that no fundamental background material on single-input single-output systems and linear algebra is presented in the present volume, unlike in the reference given above. This is because we feel that such material is readily available from many sources and in particular the quoted reference may be consulted.

The organization of a vacation school and at the same time the preparation of a book to be published before the school takes place is a daunting task. We certainly would not have succeeded without the full cooperation of all the lecturers involved in the school, the typist who prepared the camera-ready copy and the staff of the publishers. Our grateful thanks therefore go to all those who honoured deadlines and delivered the drafts of the chapters which follow this preface, to Vera Butterworth who typed the whole volume under exceptionally trying circumstances, and to Pauline Maliphant of Peter Peregrinus Ltd. who waited so patiently for the chapters to arrive at Stevenage from Manchester. This book would not have appeared but for the support given by the Science and Engineering Research Council and our sincere thanks go to staff of the SERC and to the other members of the planning panel which organized the vacation school.

D.J. Bell

P.A. Cook

N. Munro

Manchester, 1982.

Chapter 1

State-space theory

Dr. P. A. COOK

Synopsis

A brief review is given of the use of state-space methods in the study of linear time-invariant systems. Topics covered include modal decomposition, controllability and observability, standard forms, calculation of transfer function matrices, construction of state-space realizations and system inversion.

1.1 Introduction

We shall be concerned with linear stationary dynamical systems described by equations of the general form

$$\dot{x} = Ax + Bu \tag{1.1}$$

$$y = Cx + Du \tag{1.2}$$

where the vectors u (inputs), x (states) and y (outputs) have respective dimensions m,n,p, and A,B,C,D are constant matrices. Usually we shall have D = 0.

Solving equation (1.1) with initial conditions

$$x(0) = x_0 \tag{1.3}$$

and substituting into equation (1.2), we obtain

$$y(t) = C\exp(At)x_0 + \int_0^t C\exp\{A(t-\tau)\}Bu(\tau)d\tau + Du(t) \tag{1.4}$$

which, after Laplace transformation, gives

$$\bar{y}(s) = C(sI-A)^{-1}x_0 + G(s)\bar{u}(s)$$

where

$$G(s) = C(sI-A)^{-1}B + D \tag{1.5}$$

is the transfer function matrix.

Clearly, $G(s)$ is unaltered by coordinate transformation in the state space, i.e.

$$A \to R^{-1}AR, \quad B \to R^{-1}B, \quad C \to CR, \quad D \to D$$

where R is a nonsingular constant matrix. In particular, if A has distinct eigenvalues $(\lambda_1,\ldots,\lambda_n)$, with corresponding eigenvectors $(w_1,\ldots,w_n)$ we can take

$$R = W = [w_1,\ldots,w_n]$$

so that

$$W^{-1}AW = \text{diag}\,(\lambda_1,\ldots,\lambda_n) \tag{1.6}$$

and $G(s)$ has the modal decomposition

$$G(s) = \sum_{i=1}^{n} \frac{\gamma_i\beta_i^T}{s-\lambda_i} + D \tag{1.7}$$

where λ_i is the ith column of CW and β_i^T is the ith row of $W^{-1}B$.

Although we are specifically considering continuous-time systems, the same algebraic structure applies to the discrete time case, where we have difference equations instead of differential equations. Thus, if the system is described by

$$x(t+1) = Ax(t) + Bu(t) \tag{1.8}$$

$$y(t) = Cx(t) + Du(t) \tag{1.9}$$

and we define the z-transforms by

$$\tilde{x}(z) = \sum_{t=0}^{\infty} x(t)z^{-t}, \quad \text{etc.,}$$

then it follows that

$$\tilde{y}(z) = C(zI-A)^{-1}zx(0) + G(z)\tilde{u}(z)$$

where $G(z)$ is given by substitution of z for s in equation (1.5). The solution of equations (1.8) and (1.9) gives for $t > 0$,

$$y(t) = Du(t) + CBu(t-1) + \ldots + CA^{t-1}Bu(0) + CA^{t}x(0)$$

1.2 Controllability and Observability

A system is said to be controllable if, for any initial state, it is possible to find an input which will transfer it to the zero state in a finite time. Solving (1.1) subject to (1.3) and setting

$$x(t) = 0$$

for some specified $t \neq 0$, we get

$$x_0 = -\int_0^t \exp(-A\tau)Bu(\tau)d\tau$$

which can be solved for $u(\tau)$, for any given x_0, if and only if the rows of the matrix

$$\exp(-A\tau)B$$

are linearly independent. Since the matrix exponential can be expanded in terms of $(I,A,\ldots,A^{n-1})$ by the Cayley-Hamilton theorem, a necessary and sufficient condition for controllability is

$$\text{rank } \Phi_c = n \tag{1.10}$$

where

$$\Phi_c = [B,AB,\ldots,A^{n-1}B] \tag{1.11}$$

This is also a necessary and sufficient condition for any

state to be reachable from any other, and it plays the same rôle in the discrete-time case, as may be seen by solving (1.8).

Observability means that the state of the system at a given time can be ascertained from a knowledge of its input and output over a subsequent finite time interval. From (1.4), x_0 can be determined, given the vector functions $y(t)$ and $u(t)$ over an interval, if and only if the columns of the matrix

$$C \exp (At)$$

are linearly independent. Hence, a necessary and sufficient condition for the system to be observable is

$$\text{rank } \Phi_o = n \tag{1.12}$$

where

$$\Phi_o = [C^T, A^T C^T, \ldots, (A^T)^{n-1} C^T] \tag{1.13}$$

with the superscript T denoting transposition.

Again, the same condition applies to a discrete-time system governed by (1.8) and (1.9).

The algebraic conditions (1.10) and (1.12) are clearly very similar, in view of (1.11) and (1.13). It can be seen that neither involves the matrix D, which may be ignored for present purposes, and that a system described by the matrices (A,B,C) is controllable if and only if the "dual" system described by (A^T, C^T, B^T) is observable.

Both controllability and observability are evidently unaffected by similarity transformations in the state space. In particular, if the eigenvalues of A are all distinct, we can perform the transformation (1.6), and it follows that the rows of $\exp (At)B$ are linearly independent, i.e. the system is controllable if and only if all the vectors β_i in (1.7) are non-vanishing. Similarly, it is observable if and only if all the $\gamma_i \neq 0$ in (1.7), and so a controllable and observable system is one whose transfer function matrix contains contributions from all its modes.

In general, a system can be decomposed into four parts, characterised by their controllability and observability properties[1]. By means of a suitable coordinate transformation, the matrices (A,B,C) can be brought to the forms

$$A = \begin{bmatrix} A_{cu} & x & x & x \\ 0 & A_{co} & 0 & x \\ 0 & 0 & A_{uu} & x \\ 0 & 0 & 0 & A_{uo} \end{bmatrix} \qquad B = \begin{bmatrix} x \\ B_o \\ 0 \\ 0 \end{bmatrix}$$

$$C = [0 \quad C_c \quad 0 \quad x]$$

where the crosses denote non-vanishing blocks and the subscripts indicate controllability, observability, uncontrollability or unobservability. Only the controllable and observable part contributes to the transfer function matrix, which is now expressed as

$$G(s) = C_c(sI-A_{co})^{-1}B_o + D$$

1.3 Standard Forms

Let us suppose that the system described by (1.1) is controllable, and that the inputs enter independently of each other into the state equations, i.e.

$$\text{rank } B = m \tag{1.14}$$

Since the matrix Φ_c in (1.11) has full rank, we can construct a nonsingular square matrix by examining the columns of Φ_c in order and retaining only those which are linearly independent of their predecessors. The resulting matrix can then be rearranged in the form[2]

$$H = [b_1,\ldots,A^{\mu_1-1} b_1;\ldots;b_m,\ldots,A^{\mu_m-1} b_m] \tag{1.15}$$

where $B = [b_1,\ldots,b_m]$ and the integers $(\mu_1,\ldots,\mu_m)$ are known as the controllability indices of the system. Because of (1.14), they are all strictly positive and, since H has n columns,

$$\mu_1 + \ldots + \mu_m = n$$

Since they are determined from linear dependence relations, the controllability indices are unaltered by state-space coordinate transformations. Also, because of their relation to the ranks of the matrices $B, [B, AB], \ldots,$ they are unaffected, except for rearrangement among themselves, by non-singular transformations on the input space, i.e. recombination of inputs, and also by state-variable feedback[3]. The quantity

$$\nu_c = \max_{1 \le i \le m} \mu_i$$

is thus invariant under all these transformations. It is the smallest integer ν such that

$$\text{rank}\,[B, AB, \ldots, A^{\nu-1}B] = n$$

and is sometimes called the "controllability index".

We now perform the coordinate transformation

$$A \to H^{-1}AH, \quad B \to H^{-1}B$$

with H given by (1.15) .

The results have the following form: $H^{-1}B$ is a matrix with unity in positions $(1,1)$, $(\mu_1+1,2), \ldots, (\mu_1+\ldots+\mu_{m-1}+1,m)$ and zeros elsewhere, while

$$H^{-1}AH = \begin{bmatrix} A_{11} & \cdots & A_{1m} \\ \vdots & & \vdots \\ A_{m1} & \cdots & A_{mm} \end{bmatrix}$$

where the blocks have the structure

$$A_{ii} = \begin{bmatrix} 0 & \cdots & 0 & \alpha_{ii0} \\ 1 & \ddots & \vdots & \vdots \\ 0 & \ddots & 0 & \vdots \\ \vdots & \ddots & \ddots & \vdots \\ 0 \cdots 0 & & 1 & \alpha_{ii\mu_i - 1} \end{bmatrix} \qquad A_{ji} = \begin{bmatrix} 0 & \cdots & 0 & \alpha_{ij0} \\ \vdots & & \vdots & \vdots \\ 0 & \cdots & 0 & \alpha_{ii\mu_j - 1} \end{bmatrix}$$

and the entries in the final columns are the coefficients in the relations[4]

$$A^{\mu_i} b_i = \left[\sum_{j=1}^{i-1} \sum_{k=0}^{\min(\mu_i, \mu_j - 1)} + \sum_{j=i}^{m} \sum_{k=0}^{\min(\mu_i, \mu_j) - 1} \right] \alpha_{ijk} A^k b_j \tag{1.16}$$

It follows from (1.16) that the number of nonvanishing coefficients depends upon the controllability indices. The maximum possible number is nm, and this can only be achieved in the case where the first n columns of Φ_c are linearly independent, so that the indices become

$$\mu_1 = \ldots = \mu_r = \nu_c, \; \mu_{r+1} = \ldots = \mu_m = \nu_c - 1$$

where r is an integer satisfying

$$n = m(\nu_c - 1) + r$$

This is sometimes referred to as the "generic" case, since any system can be converted to it by an arbitrarily small perturbation. Consequently, it is the case which can normally be expected to occur in practice, although it does not always do so[5].

There is also another type of standard form, which can be obtained as follows. Let the rows numbered $\mu_1, \mu_1 + \mu_2, \ldots, n$, in H^{-1} be denoted by $g_1^T, \ldots, g_m^T$, respectively, and construct the matrix

$$Q = [g_1, \ldots, (A^T)^{\mu_1 - 1} g_1; \ldots; g_m, \ldots, (A^T)^{\mu_m - 1} g_m] \tag{1.17}$$

This matrix is nonsingular[2], and we can use it to perform another coordinate transformation.

We obtain the following structure:

$$Q^TA(Q^T)^{-1} = \left[\begin{array}{c|c|c}
\begin{matrix}0 & 1 & & \\ & \ddots & \ddots & \\ & & 0 & 1\end{matrix} & & \\
x \text{-------} x & x \text{-------} x & x \text{-------} \\ \hline
& \begin{matrix}0 & 1 & & \\ & \ddots & \ddots & \\ & & 0 & 1\end{matrix} & \\
x \text{-------} x & x \text{-------} x & x \text{-------} \\ \hline
& & \begin{matrix}0 & 1 & \\ & \ddots & \ddots \end{matrix}
\end{array}\right],$$

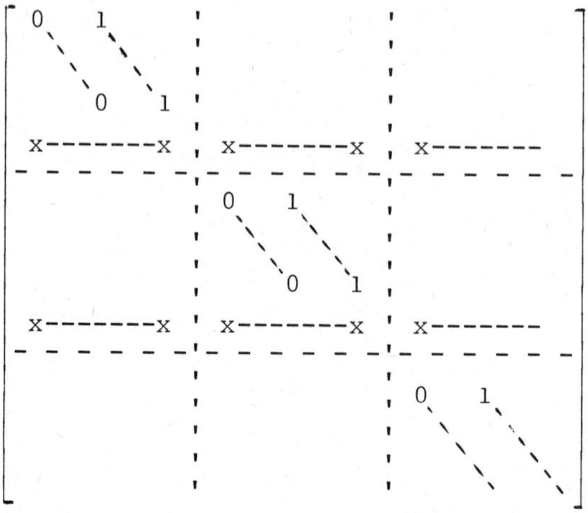

$$Q^TB = \left[\begin{array}{ccc}
0 & & \\
\vdots & & \\
0 & & \\
1 & x \text{-------} & \\ \hline
& 0 & \\
& \vdots & \\
& 0 & \\
& 1 & x \text{---} \\ \hline
& & 0 \\
& & \vdots
\end{array}\right]$$

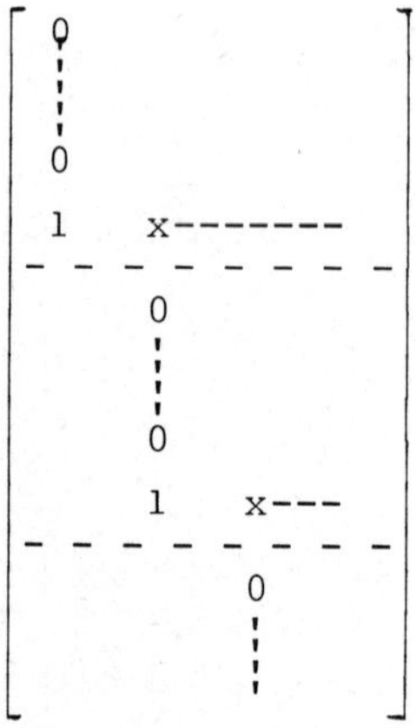

where the crosses denote nonzero elements. By further similarity transformations, the number of nonvanishing entries can be reduced to the same as that in the previous standard form. The block sizes are also the same as in the previous form.

In view of the analogy between (1.11) and (1.13), we can obtain similar standard forms for A^T and C^T in the case of an observable system. The block sizes are then given by a set of integers $(\pi_1,\ldots,\pi_p)$, obtained from the columns of Φ_o and called observability indices. These are strictly positive provided that the outputs in (1.2) are independent functions of the states, i.e.

$$\text{rank } C = p$$

and they satisfy

$$\pi_1 + \ldots + \pi_p = n$$

The maximum among them,

$$\nu_o = \max_{1 \leq i \leq p} \pi_i$$

is sometimes called the "observability index".

1.4 Calculation of the Transfer Function Matrix

We now consider the problem of computing $G(s)$, as given by (1.5), from the matrices (A,B,C,D). If a modal decomposition is possible, the result can be obtained by summing the terms in (1.7), but this is a roundabout method because of the need to calculate the eigenvalues and eigenvectors of A.

A more direct method is to compute the resolvent matrix $(sI-A)^{-1}$ and then multiply it by B and C. An algorithm which is often used is variously attributed to Souriau, Frame and Faddeev[6], although its essential features were discovered by Leverrier[7]. It calculates the coefficients in the expansions

$$\text{adj } (sI-A) = R_{n-1}s^{n-1} + \ldots + R_1 s + R_0$$

and

$$\det (sI-A) = s^n + a_{n-1}s^{n-1} + \ldots + a_1 s + a_0$$

as follows: since

$$(sI-A) \text{ adj } (sI-A) = I \det (sI-A)$$

we have, by equating coefficients,

$$
\begin{aligned}
R_{n-1} &= I \\
R_{n-2} &= AR_{n-1} + a_{n-1}I \\
&\vdots \\
R_0 &= AR_1 + a_1I \\
0 &= AR_0 + a_0I
\end{aligned}
\tag{1.18}
$$

whence, taking traces,

$$
\begin{aligned}
\mathrm{tr}\, R_{n-2} &= \mathrm{tr}\, (AR_{n-1}) + n\, a_{n-1} \\
&\vdots \\
\mathrm{tr}\, R_0 &= \mathrm{tr}\, (AR_1) + n\, a_1 \\
0 &= \mathrm{tr}\, (AR_0) + n\, a_0
\end{aligned}
\tag{1.19}
$$

while, from the identity

$$\frac{d}{ds} \det (sI-A) = \mathrm{tr}\, \mathrm{adj}\, (sI-A)$$

we get

$$
\begin{aligned}
(n-1)\, a_{n-1} &= \mathrm{tr}\, R_{n-2} \\
&\vdots \\
a_1 &= \mathrm{tr}\, R_0
\end{aligned}
$$

which, together with (1.19), yields

$$
\begin{aligned}
a_{n-1} &= -\, \mathrm{tr}\, (AR_{n-1}) \\
&\vdots \\
a_1 &= -\, \frac{1}{n-1}\, \mathrm{tr}\, (AR_1) \\
a_0 &= -\, \frac{1}{n}\, \mathrm{tr}\, (AR_0)
\end{aligned}
\tag{1.20}
$$

The sets of equations (1.18) and (1.20) provide an iterative computational scheme, with the last equation of (1.18) serving as a check on the procedure. This is desirable, since the method has been found to be numerically inaccurate in some cases[8].

The number of multiplications required by this algorithm is $O(n^4)$. Several methods have been proposed which can be computationally more efficient than this[9]. In general, they involve the conversion of A, by a similarity transformation, into a companion form

$$\begin{bmatrix} 0 & \cdots & \cdots & 0 & -a_0 \\ 1 & \ddots & & \vdots & \vdots \\ 0 & \ddots & \ddots & \vdots & \vdots \\ \vdots & \ddots & \ddots & 0 & \vdots \\ 0 & \cdots & 0 & 1 & -a_{n-1} \end{bmatrix} \quad \text{or} \quad \begin{bmatrix} 0 & 1 & 0 & \cdots & 0 \\ \vdots & \ddots & \ddots & \ddots & \vdots \\ \vdots & & \ddots & \ddots & 0 \\ 0 & \cdots & \cdots & 0 & 1 \\ -a_0 & \cdots & \cdots & \cdots & -a_{n-1} \end{bmatrix}$$

from which det (sI-A) and adj (sI-A) may readily be obtained. The transformation can be performed with $O(n^3)$ multiplications and divisions, but it is also subject to numerical difficulties and there does not seem to be any convenient way of checking the results.

1.5 Realizations

Next, we consider the converse problem of constructing a state-space realization, i.e. a set of matrices (A,B,C) which give rise, through (1.5), to a specified transfer function matrix. We assume that G(s) is strictly proper, i.e. $G(s) \to 0$ as $s \to \infty$, so that $D = 0$. There are many ways of setting up a realization; one convenient systematic procedure is the following.

First, express each column of G(s) as a polynomial vector divided by a common denominator. Specifically, let the least common denominator of the ith column be

$$s^{n_i} + d_{in_i-1}s^{n_i-1} + \ldots + d_{i0}$$

and the corresponding numerator vector be

$$q_{in_i-1}s^{n_i-1} + \ldots + q_{i0}$$

Then, a realization can be immediately be written down in the form

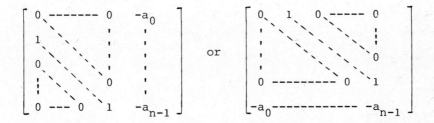

$$A = \begin{bmatrix} A_1 & & 0 \\ & \ddots & \\ 0 & & A_m \end{bmatrix}, \quad B = \begin{bmatrix} B_1 & & 0 \\ & \ddots & \\ 0 & & B_m \end{bmatrix}, \quad C = [C_1 \ldots C_m]$$

where

$$A_i = \begin{bmatrix} 0 & 1 & 0 & \cdots & 0 \\ \vdots & \ddots & \ddots & \ddots & \vdots \\ \vdots & & \ddots & \ddots & 0 \\ 0 & \cdots & \cdots & 0 & 1 \\ -d_{i0} & \cdots & \cdots & \cdots & -d_{in_i-1} \end{bmatrix}, \quad B_i = \begin{bmatrix} 0 \\ \vdots \\ \vdots \\ 0 \\ 1 \end{bmatrix}$$

and $$C_i = [q_{i0}, \ldots, q_{in_i-1}]$$

This realization is controllable, by construction, but not necessarily observable. However, it can be reduced to an observable form, without losing controllability, by performing a similarity transformation, constructed out of elementary row and column operations[10]. An alternative realization could, of course, be generated by starting with the rows, instead of the columns, of $G(s)$. It would then automatically be observable but might need to be reduced to controllable form.

In the discrete-time case, it may be convenient to start from the "Markov parameters"

$$K_i = CA^{i-1}B$$

which appear in the solution of (1.8) and (1.9), and which are also the coefficient matrices in the Laurent expansion

$$G(z) = D + \sum_{i=1}^{\infty} K_i z^{-i}$$

We define the matrices

$$K = \begin{bmatrix} K_1 & K_2 & \cdots & K_n \\ K_2 & & & \vdots \\ \vdots & & & \vdots \\ K_n & \cdots & & K_{2n-1} \end{bmatrix}, \quad \tilde{K} = \begin{bmatrix} K_2 & K_3 & \cdots & K_{n+1} \\ K_3 & & & \vdots \\ \vdots & & & \vdots \\ K_{n+1} & \cdots & & K_{2n} \end{bmatrix}$$

and then, from (1.11) and (1.13),

$$K = \Phi_o^T \Phi_c , \quad \tilde{K} = \Phi_o^T A \Phi_c \tag{1.21}$$

so that, if the matrices (A,B,C) give a controllable and observable realization,

$$\text{rank } K = n$$

and hence two nonsingular matrices, M and N, can be found such that

$$MKN = \begin{bmatrix} I_n & 0 \\ 0 & 0 \end{bmatrix} \tag{1.22}$$

by using, for example, Gauss reduction[10]. From (1.21) and (1.22), it follows that, for some nonsingular matrix L,

$$M \Phi_o^T = \begin{bmatrix} L \\ x \end{bmatrix}, \quad \Phi_c N = [L^{-1} \quad x]$$

where the crosses denote irrelevant blocks, and hence

$$MK = \begin{bmatrix} LB & x \\ x & x \end{bmatrix}, \quad KN = \begin{bmatrix} CL^{-1} & x \\ x & x \end{bmatrix}$$

and

$$M\tilde{K}N = \begin{bmatrix} LAL^{-1} & x \\ x & x \end{bmatrix}$$

from which the controllable and observable realization (LAL^{-1}, LB, CL^{-1}) can immediately be extracted[11].

1.6 Inverse Systems

From a state-space viewpoint, the problem of inverting a rational function matrix G(s) is equivalent to finding a realization of a transfer function matrix F(s) such that

$$F(s)G(s) = s^{-q} I_m$$

for a left inverse, or

$$G(s)F(s) = s^{-q} I_p$$

for a right inverse, where q is a non-negative integer, assuming that F(s) and G(s) are both proper, i.e. bounded as $s \to \infty$. It is useful to define the system matrix[10]

$$P(s) = \begin{bmatrix} sI-A & B \\ -C & D \end{bmatrix}$$

since a left or right inverse exists if and only if the normal rank of P(s) is (m+n) or (p+n), respectively[12]. We are assuming that the realization (A,B,C) is controllable and observable since we are concerned only with properties which affect G(s). The values of s for which P(s) has less than full rank are sometimes known as the transmission zeros[13] of G(s).

If an inverse exists, it may be computed in various ways. One method[12] is to perform repeated manipulations on the system until the D matrix becomes of full rank, when the inverse can be constructed directly. If D is square and nonsingular, in fact, a realization of the unique inverse $G^{-1}(s)$ can immediately be obtained from (1.1) and (1.2) in the form

$$\dot{x} = (A - BD^{-1}C)x + BD^{-1}y \qquad (1.23)$$

$$u = -D^{-1}Cx + D^{-1}y \qquad (1.24)$$

with u regarded as output and y as input. For a square

system, i.e. $p = m$, the inverse is necessarily unique and two-sided if it exists at all, but (1.23) and (1.24) cannot be be used to construct it unless D is nonsingular, whereas normally, on the contrary, D is null.

However, the inverse of a square system can also be expressed as

$$G^{-1}(s) = [0 \quad I_m]P^{-1}(s)\begin{bmatrix} 0 \\ I_m \end{bmatrix} \tag{1.25}$$

and, by suitable elementary transformations[14], $P(s)$ can be reduced to the form

$$\begin{bmatrix} sI-\tilde{A} & 0 \\ 0 & I+sJ \end{bmatrix}$$

where J is a Jordan matrix with all its eigenvalues equal to zero. The inverse matrix (1.25) then consists of two parts, one of which depends on $(sI-\tilde{A})^{-1}$ and is strictly proper, while the other contains $(I + sJ)^{-1}$ and is a polynomial matrix.

References

1. KALMAN, R.E. : 'Canonical structure of linear dynamical systems', Proc. Nat. Acad. Sci., 48, (1962), 596-600.
2. LUENBERGER, D.G. : 'Canonical forms for linear multivariable systems', IEEE Trans. Autom. Control, AC-12, (1967), 290-293.
3. BRUNOVSKY, P. : 'A classification of linear controllable systems', Kybernetica, 3, (1970), 173-187.
4. POPOV, V.M. : 'Invariant description of linear time-invariant systems', SIAM J. Control, 10, (1972), 252-264.
5. COOK, P.A. : 'On some questions concerning controllability and observability indices', Proc. 7th IFAC Triennial World Congress (Helsinki, 1978), pp.1699-1705.
6. FADDEEV, D.K. and V.N. FADDEEVA : Computational Methods of Linear Algebra , (Freeman 1963).
7. BARNETT, S. : Introduction to Mathematical Control Theory , (Oxford University Press 1975).
8. BOSLEY, M.J. H.W. KROPHOLLER, F.P. LEES and R.M. NEALE, 'The determination of transfer functions from state-

variable models', Automatica, 8, (1972), 213-218.

9. DALY, K.C., : 'A computational procedure for transfer-function evaluation', Int. J. Control, 20, (1974), 569-576.

10. ROSENBROCK, H.H. : State-Space and Multivariable Theory , (Nelson-Wiley 1970).

11. HO, B.L. and R.E. KALMAN, : 'Effective construction of linear state-variable models from input-output functions', Regelungstechnik, 14, (1966), 545-548.

12. MOYLAN, P.J. : 'Stable inversion of linear systems', IEEE Trans. Autom. Control, AC-22, (1977), 74-78.

13. DAVISON, E.J. and S.H. WANG : 'Properties and calculation of transmission zeros of linear multivariable systems', Automatica, 10, (1974), 643-658.

14. VAN DER WEIDEN, A.J.J. : 'Inversions of rational matrices', Int. J. Control, 25, (1977), 393-402.

Problems

P.1.1 A system is described by the matrices

$$A = \begin{bmatrix} -2 & 1 & 1 & 0 \\ -1 & 0 & 0 & -1 \\ -1 & 0 & -1 & -1 \\ 0 & 1 & 1 & -1 \end{bmatrix}, \quad B = \begin{bmatrix} 1 & 0 \\ 0 & 1 \\ 0 & 1 \\ 1 & 0 \end{bmatrix}$$

$$C = \begin{bmatrix} 0 & -1 & -1 & 0 \\ 1 & 0 & 0 & 1 \end{bmatrix}, \quad D = \begin{bmatrix} 0 & 0 \\ 0 & 0 \end{bmatrix}$$

Convert the system to standard form with the help of the matrix Q given by equation (1.17) .

P.1.2 For the above system obtain the resolvent matrix $(sI-A)^{-1}$ from the recurrence relations (1.18) and (1.20), and hence calculate $G(s)$.

Chapter 2

Complex variable methods in feedback systems analysis and design

Professor A. G. J. MACFARLANE

Synopsis

Generalized Nyquist and generalized Root-Locus diagrams are algebraic curves derived from the spectral analysis of appropriate matrix-valued functions of a complex variable. Their key properties can be comprehensively analysed in terms of a state-space model of the feedback loop which is being studied. Such an analysis is essentially geometrical in nature and is concerned with the way in which certain subspaces, defined via the various operators involved, sit with respect to one another, and with ways of assigning complex frequencies to these subspaces. The current status of work on the design of feedback systems using these results is briefly discussed, and possible future developments are noted.

2.1 Introduction

The original Nyquist stability criterion owed its central importance in classical feedback control work to the fact that it tested the stability of a proposed feedback loop in terms of a directly *measured* transmission characteristic. This radical departure of inferring stability from measured signal characteristics instead of from computed dynamical-model characteristics placed complex-function methods in a centrally-important position in the development of control analysis and design techniques up to the late 1940s. The root-locus technique developed in the 1950s added further weight to the arguments for using complex-variable approaches in the analysis and design of feedback systems. At the time of the introduction of the root-locus method, the close link with the Nyquist technique was not appreciated; one of the virtues of the algebraic function approach used here to generalize these techniques to the multivariable case is that it shows how both these classical techniques are essentially different ways of representing curves derived from a single algebraic function relating complex frequency and complex gain.

Following the widespread introduction of state-space methods in the 1960s, interest in complex-variable approaches to feedback systems analysis and design declined until the advent of computer-driven interactive graphic displays reawakened interest in "picture-generating" techniques. The work presented here arose out of a desire to do two things:

(i) to generalize both the Nyquist-Bode and Root-Locus techniques to the multivariable case; and
(ii) to relate these generalizations to state-space descriptions.

As such it is directed towards a fusing of state-space and classical frequency-response ideas, and to an exploration of the way in which the emergence of computer graphics can be best exploited in the further development of multivariable feedback system design techniques.

2.2 Generalized Nyquist and Root-Locus Diagrams

The dual role of complex numbers is familiar from elementary complex number theory; they can be used to represent both objects (e.g. vectors in a plane) and operations on objects (e.g. rotation and stretching of vectors in a plane). This dual role is important in classical single-input single output frequency-response methods, where complex numbers are used to represent *frequencies* (complex numbers) and *gains* (operations on complex vectors). The two classical approaches to the complex-variable-based analysis of feedback systems:

(i) study open-loop gain as a function of frequency (the

Nyquist-Bode approach); or

(ii) study closed-loop frequency as a function of gain (the Evans Root-Locus approach).

Our first objective is to show that there is an intimate relationship between the generalization of these two approaches to the multivariable (many-input many-output) case and the standard state-space model

$$\Sigma : \quad \dot{x} = Ax + Bu \qquad\qquad y = Cx + Du \tag{2.1}$$

In most of what follows the matrix D will be taken to be zero for simplicity of exposition. However, its inclusion at this point gives completeness and symmetry to the development of the general case. [A is $n\times n$; B is $n\times \ell$; C is $m\times n$; D is $m\times \ell$. When vector feedback loops are being discussed $\ell = m$]. Suppose we sever m feedback connections to an arbitrary linear dynamical system as shown in Fig. 2.1, and that the transmittance between the injection point α and the return α' is given by a state-space model of the form shown in equation (2.1). Then on remaking the feedback connections, we see that the *closed-loop* characteristic frequencies of this system will be given by the spectrum (set of eigenvalues) of the matrix

$$S = A + B(I_m - D)^{-1}C \tag{2.2}$$

and that the *open-loop* transfer function matrix relating vector exponential signals injected at α to the vector exponential signals returned at α' is

$$G(s) = D + C(sI_n - A)^{-1}B \tag{2.3}$$

(I_m and I_n are unit matrices of orders m and n respectively.) If one compares the right-hand sides of (2.2) and (2.3) they are seen to have a suggestive structural similarity.

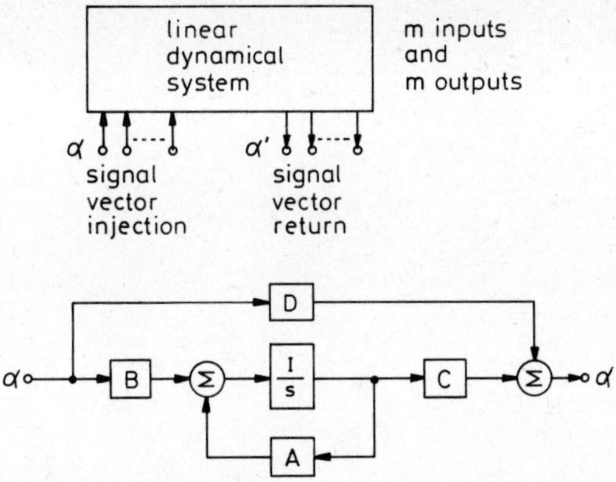

Fig 2.1

This becomes an exact structural equivalence if we introduce a variable g and change (2.2) to the form

$$S(g) \;=\; A + B(gI_m - D)^{-1}C \qquad (2.4)$$

where the role of the variable g will emerge shortly. Our aim is to put, in a state-space-model context, the roles of open-loop gain (the key Nyquist-Bode concept) and closed-loop frequency (the key Evans Root-Locus concept) *on an exactly equal footing* as complex variables. With this in mind we look for a system giving the closed-loop frequency matrix of equation (2.4) and find that it is as shown in Fig. 2.2, where g is now interpreted as a gain parameter. On redrawing Fig. 2.2 in the form shown in Fig. 2.3 we now see that the variables s and g have indeed been placed in a strikingly symmetrical relationship.

In Fig. 2.3 s is to be interpreted as a complex frequency variable and g as a complex gain variable. For the related vector feedback loop (with m inputs and m outputs) the open-loop matrix

$$G(s) \;=\; D + C(sI_n - A)^{-1}B$$

describes open-loop gain as a function of imposed frequency

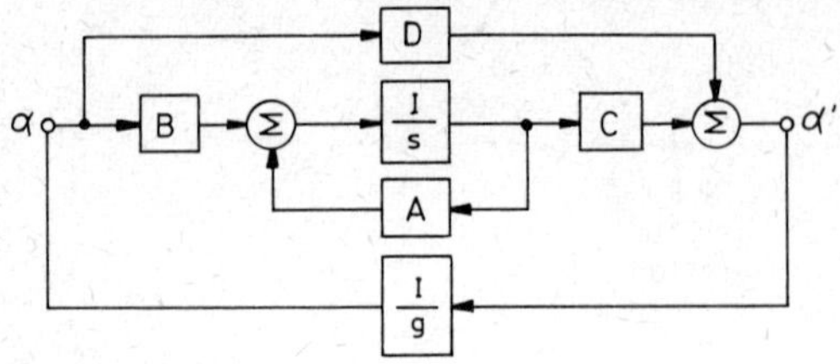

Fig 2.2

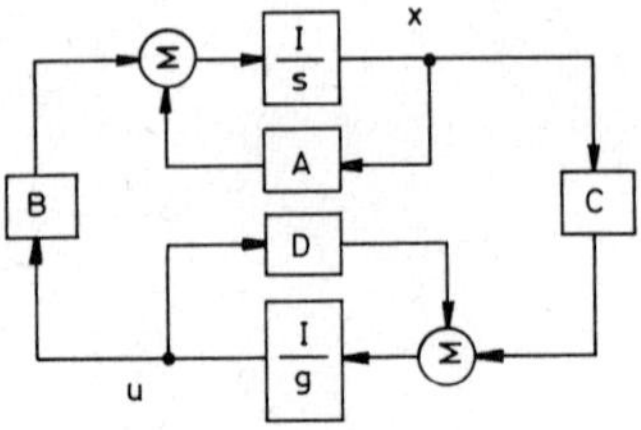

Fig 2.3

s; and the closed-loop frequency matrix

$$S(g) = A + B(gI_m - D)^{-1}C$$

describes the closed-loop frequency as a function of the imposed gain parameter g.

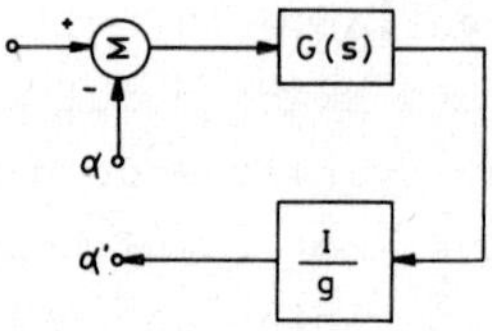

Fig. 2.4

For the negative feedback arrangement of Fig. 2.4, if

$$F(s) = I_m + g^{-1}G(s)$$

is the return difference matrix for the set of feedback loops broken at $\alpha-\alpha'$ then one can show that[1]

$$\frac{\det[sI_n - S(g)]}{\det[sI_n - S(\infty)]} = \frac{\det[gI_m - G(s)]}{\det[gI_m - G(\infty)]} = \frac{\det F(s)}{\det F(\infty)} = \frac{\mathrm{CLCP}(s)}{\mathrm{OLCP}(s)} \tag{2.5}$$

where CLCP(s) and OLCP(s) are the system's closed-loop and open-loop characteristic polynomials respectively. The importance of equation (2.5) is that it tells us that in this archetypal negative feedback situation we may equally well study the effect of closing feedback loops on stability in terms of an open-loop gain description or a closed-loop frequency description. Specifically it shows that, for values of s not in the spectrum of A and values of g not in the spectrum of D

$$\det[sI_n - S(g)] = 0 \iff \det[gI_m - G(s)] = 0$$

This in effect says that a knowledge of how the characteristic values of $G(s)$ vary as a function of the frequency parameter s is equivalent (for the purposes of making inferences about closed-loop stability) to a knowledge of how the characteristic values of $S(g)$ vary as a function of the gain parameter g.

2.2.1 Characteristic frequencies and characteristic gains

For a given value of the gain parameter g the eigenvalues $\{s_i : i = 1,2,\ldots,n\}$ of $S(g)$ are the corresponding set of closed-loop characteristic frequencies, obtained from

$$\det[sI_n - S(g)] = 0$$

For a given value of the frequency parameter s the eigenvalues $\{g_i : i = 1,2,\ldots,m\}$ of $G(s)$ may be called the corresponding set of open-loop *characteristic gains*, obtained from

$$\det[gI_m - G(s)] = 0$$

The closed-loop characteristic frequencies $\{s_i\}$ for a given value of g are, as is well known, associated with invariant subspaces in the state space. The open-loop characteristic gains $\{g_i\}$ for a given value of s are associated with invariant subspaces in an input-output space.

From the characteristic equations for $G(s)$ and $S(g)$, namely

$$\Delta(g,s) \triangleq \det[gI_m - G(s)] = 0$$

and $$\nabla(s,g) \triangleq \det[sI_n - S(g)] = 0$$

one obtains a pair of algebraic equations relating the complex variables s and g. These define a pair of algebraic functions[1]:

(i) a characteristic gain function $g(s)$ which gives open-loop characteristic gain as a function of frequency;

and

(ii) a characteristic frequency function $s(g)$ which gives closed-loop characteristic frequency as a function of gain.

The importance of these two algebraic functions lies in the fact that they are the natural means of generalizing the concepts of Nyquist diagram and Root-Locus diagram to the multivariable case[1].

2.2.2 Generalized Nyquist Diagrams and the Generalized Nyquist Stability Criterion

The characteristic gain loci (generalized Nyquist diagrams) for the m-vector feedback loop with transmittance matrix $G(s)$ are the loci in the complex gain plane which are traced out by the eigenvalues of $G(s)$ as s transverses the so-called Nyquist D-contour in the complex frequency plane (s-plane). They can also be defined as the $\pm 90^{\circ}$ constant-phase contours on the g-plane of the algebraic function $s(g)$. Their utility lies in their role in the following sort of generalization of the Nyquist Stability Criterion[1,2].

Multivariable Nyquist Stability Criterion: The multivariable Nyquist stability criterion relates the closed-loop stability of the configuration of Fig. 2.5 to the characteristic gain loci for the loop-gain matrix $G(s)H(s)$. Suppose the system of Fig. 2.5 has no uncontrollable and/or unobservable modes whose corresponding characteristic frequencies lie in the right-half s-plane. Then this feedback configuration will be closed-loop stable if and only if the net sum of anticlockwise

Fig 2.5

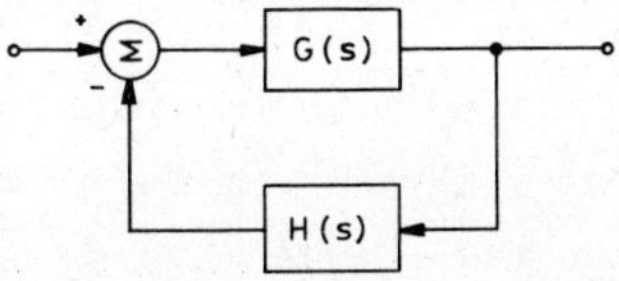

encirclements of the critical point (-1+j0) by the set of characteristic gain loci of G(s)H(s) is equal to the total number of right-half-plane poles of G(s) and H(s).

Proofs of multivariable versions of the Nyquist Stability Criterion have been given by Barman and Katzenelson[3] and MacFarlane and Postlethwaite[2], and Postlethwaite and MacFarlane[1] following earlier heuristic approaches by Bohn and Kasavand[4] and MacFarlane[5]. More general versions have been considered by Harris and Valenca[52] and by Desoer and Wang[53].

2.2.3 Multivariable Root Loci

The loci of open-loop characteristic gain with frequency are generalizations of classical Nyquist diagrams in the complex gain plane; there is a corresponding generalization of the classical Evans' Root Locus diagram in the frequency plane. These are the characteristic frequency loci which are traced out by the eigenvalues of S(g) as g traverses the negative real axis in the gain plane. They can also be regarded as the 180° phase contours of g(s) on the frequency plane. The theory behind the multivariable root locus has been given by Postlethwaite and MacFarlane[1] and their use discussed by Retallack[6], Kouvaritakis and Shaked[7], MacFarlane, Kouvaritakis and Edmunds[8], Kouvaritakis and Edmunds[9], Kouvaritakis[10] and Owens[11]. Many examples of generalized Nyquist and Root-Locus diagrams are given in ref.12; a complete review of the development of frequency-response methods is given in ref. 13.

2.2.4 Conformal Nature of Mapping Between Frequency and Gain

The algebraic functions g(s), giving characteristic gain

as a function of frequency, and $s(g)$, giving characteristic frequency as a function of gain contain essentially the same information. One is simply a "re-packaged" version of the other. One would thus expect there to be an exact structural relationship between the (generalized) Nyquist and (generalized) Root Locus diagrams for a given feedback loop. These structural relationships also reflect the fact that the mappings between s-plane and g-plane and vice-versa are conformal, so that appropriate angular relationships are preserved. In correlating the main features of the two forms of diagram the following rule has been found useful.

Locus-Crossing Rule: Let D be a path in the s-plane and let D' be its image under $g(s)$ in the g-plane. Also let E be a path in the g-plane and let E' be its image under $s(g)$ in the s-plane. Then the following relationships hold.

(i) Each crossing of E by D' corresponds to a related crossing of D by E', and vice-versa.

(ii) If E' crossed D from left to right with respect to a given orientation for D, then E will cross D' from left to right with respect to the induced orientation for D', and vice-versa. (The given orientations for D and E are taken to induce appropriate orientations in D' and E' : that is a traversal of D by s in a positive orientation, makes $g(s)$ traverse D' in a positive orientation, with a similar convention for E and E'.)

(iii) The angle at which E' crosses D is the same as the angle at which D' crosses E.

This rule is illustrated in Fig. 2.6 taking D as the positive imaginary axis in the s-plane and E as the negative real axis in the g-plane. This shows that a crossing of the negative real axis in the gain plane (at g' say) by a portion of the generalized Nyquist diagram corresponds to the migration of a closed-loop characteristic frequency from left to right (under negative feedback of amount $(g')^{-1}$ applied equally in all loops).

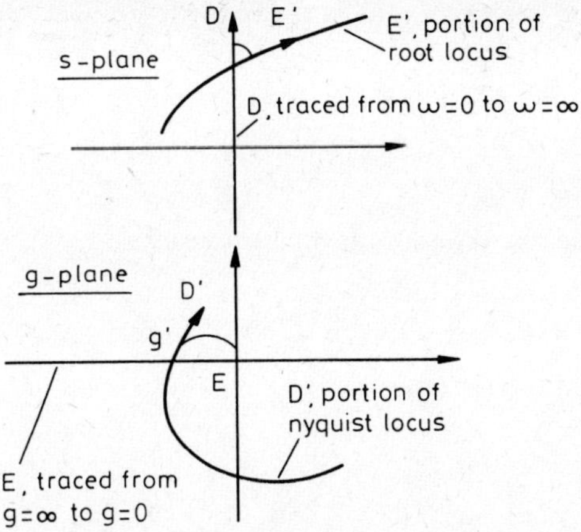

Fig. 2.6

2.3 Zeros

In physical terms zeros are associated with the vanishing of vector gain; that is with the existence of non-zero input exponential signal vectors which result in zero output[23]. The discussion of zeros in the literature is confusing because one can talk about zeros in connection with various objects: $\Sigma(A,B,C,D)$, or $G(s)$, or $g(s)$ for example. The object $\Sigma(A,B,C,D)$ has associated with it a larger set of zeros than the object $G(s)$ because in passing from the representation $\Sigma(A,B,C,D)$ to the representation $G(s)$ one discards phenomena (uncontrollability and unobservability) which can be discussed in terms of zeros (the so-called decoupling zeros[15]). In turn the object $G(s)$ has a larger set of zeros than the object $g(s)$ because vanishing of all characteristic gains does not necessarily imply the vanishing of the vector gain for $G(s)$. For the purposes of this paper we will take the system representations being discussed to be completely controllable and completely observable, and the zeros to be what are usually called the transmission zeros[21]. A useful physical interpretation of a zero then comes from the following result.

Transmission-blocking theorem: For a system $\Sigma(A,B,C,D)$ having a number of inputs less than or equal to the number of outputs, necessary and sufficient conditions for an input of the form

$$u(t) = 1(t)e^{zt}\, u_z$$

to result in a state-space trajectory

$$x(t) = 1(t)e^{zt}\, x_2 \qquad \text{for} \quad t > 0$$

and an output

$$y(t) = 0 \qquad \text{for} \quad t > 0$$

are that a complex frequency z and a complex vector u_z exist such that

$$\begin{bmatrix} zI - A & -B \\ -C & -D \end{bmatrix} \begin{bmatrix} x_z \\ u_z \end{bmatrix} = 0 \qquad (2.6)$$

(Here $1(t) = 0$ for $t \leq 0$ and $1(t) = 1$ for $t > 0$.) The corresponding vectors u_z and x_z are called the zero directions in the input and state spaces respectively[21].

<u>Zero Pencil</u>: From (2.6) we have that, for the case when $D = 0$,

$$\begin{aligned} (zI-A)x_z &= Bu_z \\ Cx_z &= 0 \end{aligned} \qquad (2.7)$$

Zero directions in the state space naturally lie in the kernel of C so that

$$x_z = Mv_z \qquad (2.8)$$

where M is a basis matrix representation of $\operatorname{Ker} C$ and v_z is an appropriate vector of constants. Substitution of (2.8) into (2.7) and premultiplication by the full-rank transformation matrix

$$\begin{pmatrix} N \\ B^+ \end{pmatrix}$$

where N is a full-rank left annihilator of B (i.e. $NB = 0$) and B^+ is a left inverse of B (i.e. $B^+B = I_\ell$) yields

$$(z\,NM - NAM)v_z = 0$$
$$u_z = B^+(zI - A)Mv_z$$

The object $(sNM-NAM)$ is called the *zero pencil*[23,24] for the system $\sum(A,B,C)$. The zeros are the roots of the invariant factors of the zero pencil, and this fact can be used to exhibit the geometric characterisation of zeros in a very succinct way. To do this it is useful to use Sastry and Desoer's notion of restricting an operator in domain and range[26].

2.3.1 Restriction of an operator in domain and range

Given a linear map A from $\mathbb{C}^n$ to $\mathbb{C}^n$ where $\mathbb{C}^n$ is the space of n-dimensional complex vectors, and two subspaces $\mathcal{S}_1$, $\mathcal{S}_2$ of $\mathbb{C}^n$, then the restriction of A to $\mathcal{S}_1$ in the domain and $\mathcal{S}_2$ in the range is defined to be the linear map which associates with $x_1 \in \mathcal{S}_1$ the orthogonal projection of Ax_1 on to $\mathcal{S}_2$. This restricted linear map (and also its matrix representation) will be denoted by

$$A\Big|_{\mathcal{S}_1 \to \mathcal{S}_2} \triangleq \check{A}$$

If the columns of S_1 and S_2 respectively form orthonormal bases for $\mathcal{S}_1$ and $\mathcal{S}_2$ then the matrix representation of the restricted operator is given by

$$\check{A} = S_2^* A S_1$$

where * denotes complex conjugate transposition. With a mild abuse of language one then calls the roots of

$$\det[S_2^* S_1 - S_2^* A S_1] = 0$$

the spectrum of the restricted operator $\check{A}$.

2.3.2 Spectral characterisation of zeros

Suppose we choose N^* and M so that their columns are orthonormal bases for $\text{Ker } B^*$ and $\text{Ker } C$ respectively. Then

$$NM = I_n \Big|_{\text{Ker } C \to \text{Ker } B^*}$$

and

$$NAM = A \Big|_{\text{Ker } C \to \text{Ker } B^*}$$

and one may say that: the finite zeros of $\sum(A,B,C)$ are the spectrum of the restriction of A to $\text{Ker } C$ in domain and $\text{Ker } B^*$ in range.

2.3.3 Finite zeros when D is non-zero

Suppose the direct-coupling operator D for $\sum(A,B,C,D)$ having m inputs and m outputs has nullity d_o . Then $\sum(A,B,C,D)$ has at most $(n-d_o)$ finite zeros and at least d_o root-locus asymtotes going to infinity (to infinite zeros)[8]. Let D have a characteristic decomposition

$$D = [U_o \; M_o] \begin{pmatrix} J_o & 0 \\ 0 & 0_{d_o} \end{pmatrix} \begin{pmatrix} V_o \\ N_o \end{pmatrix}$$

where $[U_o \; M_o]$ and $[V_o^T \; N_o^T]^T$ are the eigenvector and dual eigenvector matrices for D and J_o is the Jordan block associated with the non-zero eigenvalues of D. Further, suppose that there is a complete set of eigenvectors spanning the null space of D. Define A_o, B_o, C_o by

$$A_o = A - BU_o J_o^{-1} V_o C$$

$$B_o = BM$$

$$C_o = N_o C$$

Then the finite zeros and root locus asymptotes of $\sum(A,B,C,D)$ are those of $\sum(A_o, B_o, C_o)$.

In case D has full rank one has that there are n finite zeros of $\sum(A,B,C,D)$ which are given by $\sigma(A-BD^{-1}C)$, the spectrum of the matrix $(A-BD^{-1}C)$.

2.4 Bilinear Transformation of Frequency and Gain Variables

If $\sum(A,B,C,D)$ having m inputs and m outputs has a D which is non-singular then, as noted above, it will have n finite zeros given by $\sigma(A-BD^{-1}C)$.

Given a bilinear transformation

$$p = \frac{a\ s + b}{c\ s + d} \tag{2.9}$$

$$s = -\frac{dp + b}{cp - a}$$

on the complex frequency it has been shown by Edmunds[28] that it is possible to find a correspondingly transformed system $\sum(\tilde{A},\tilde{B},\tilde{C},\tilde{D})$ such that

$$\tilde{G}(p) = G(s)$$

Thus, if the original system has a zero at z the transformed system will have a zero at a location given by substituting z in (2.9). One can therefore choose an appropriate bilinear transformation to get a $\tilde{D}$ of full rank, calculate the zeros as $\sigma(\tilde{A}-\tilde{B}\tilde{D}^{-1}\tilde{C})$ and transform back to find the complete set of zeros (finite and infinite) of the original system. The bilinear mapping is a conformal map between the s-plane and the p-plane for all points except $p = a/c$ which gets sent to infinity on the s-plane and $s = -d/c$ which gets sent to infinity on the p-plane. It is assumed that $(ab - bc) \neq 0$ and that the original system has no pole at $-d/c$. The set of transformed system matrices is given by:

$$\tilde{A} = (cA + dI)^{-1}(aA + bI)$$

$$\tilde{B} = (cA + dI)^{-1}B$$

$$\tilde{C} = (ad - bc)C(cA + dI)^{-1}$$

$$\tilde{D} = D - cC(cA + dI)^{-1}B$$

A similar bilinear transformation can be carried out on the complex gain variable g

$$q = \frac{a\,g + b}{c\,g + d} \qquad g = -\frac{d\,q + b}{c\,q - a}$$

The correspondingly transformed system $\sum(A',B',C',D')$ is such that

$$S'(q) = S(g)$$

and has state-space parameters given by

$$A' = A - cB(cD + dI)^{-1}C$$

$$B' = (ab - bc)B(cD + dI)^{-1}$$

$$C' = (cD + dI)^{-1}C$$

$$D' = (cD + dI)^{-1}(aD + bI)$$

2.5 Geometric Theory of Root Locus and Nyquist Diagrams

It has been noted that the finite zeros of a vector feedback loop transmittance can be characterized in terms of the spectrum of the restriction of its A-matrix in domain and range. The idea of using the spectrum of a restricted operator in connection with the Markov Parameters of a loop transfer function matrix leads to a geometric treatment of the high-gain asymptotic behaviour of generalized Root-Locus diagrams and the high-frequency asymptotic behaviour of generalized Nyquist diagrams.

Let the transfer function matrix for a strictly proper

(i.e. $D = 0$) feedback loop be expanded in a Taylor series about $s = \infty$ as

$$G(s) = \frac{G_1}{s} + \frac{G_2}{s^2} + \ldots.$$

where $(G_i : i = 1,2,\ldots.)$ are the Markov Parameters

$$G_1 = CB \qquad G_k = CA^{k-1}B \qquad k = 2,3,\ldots.$$

The use of Markov Parameters to relate the main features of generalized Root-Locus diagrams to state-space model parameters has been discussed by a number of investigators[26,29-31]. The approach and notation used here follows Sastry and Desoer[26], who give a good discussion of the assumptions involved in a detailed analysis of this sort.

Let a sequence of restricted operators be defined as follows:

$$\check{G}_k = G_k \Big|_{\text{Ker } \check{G}_{k-1} \to \text{Ker } (\check{G}_{k-1})^*}$$

and take $\check{G}_1 = G_1$.

Let d_i be the nullity (rank defect) of $\check{G}_i$. Then the high-gain asymptotic behaviour of the various branches of the generalized root-locus diagram can be determined as follows.

<u>1st Order branches</u>:

$$s_{i,1} \simeq -\frac{\lambda_{i,1}}{g} \qquad i = 1,2,\ldots(m-d_i)$$

where $\lambda_{i,1} \in \sigma[\check{G}_1]\setminus\{0\}$ and $g \to 0$

The collection of symbols $\in\sigma[\check{G}_1]\setminus\{0\}$ is to be read as "belongs to the non-zero spectrum of $\check{G}_1$ ".

<u>2nd Order branches</u>: $s^2_{i,2} \simeq -\frac{\lambda_{i,2}}{g}$ $i = 1,2,\ldots(d_2-d_1)$

where $\lambda_{i,2} \in\sigma[\check{G}_2]\setminus\{0\}$ and $g \to 0$

and for the kth order branches

$$s^k_{i,k} \simeq -\frac{\lambda_{i,k}}{g} \qquad i = 1,2,\ldots(d_k - d_{k-1})$$

where $\qquad \lambda_{i,k} \in \sigma[\check{G}_k]\setminus\{0\} \qquad$ and $\quad g \to 0$

On invoking the correspondence between the high-frequency asymptotes for the generalized Nyquist diagrams and the high-gain asymptotes for the generalized Root-Locus diagrams one then gets the following description of the high-frequency branches of the generalized Nyquist diagrams.

<u>1st Order branches</u>: $\qquad g_{i,1} \simeq -\dfrac{\lambda_{i,1}}{j\omega}$

<u>2nd Order branches</u>: $\qquad g_{i,2} \simeq -\dfrac{\lambda_{i,2}}{(j\omega)^2}$

<u>kth Order branches</u>: $\qquad g_{i,k} \simeq -\dfrac{\lambda_{i,k}}{(j\omega)^k}$

where $\qquad \lambda_{i,k} \in \sigma[\check{G}_k]\setminus\{0\}$ as before, $\quad$ and $\quad \omega \to \infty$.

For example, if the first Markov Parameter CB has full rank m then it can be shown that the state space is the direct sum of the image of the input map B and the kernel of the output map C :

$$= \text{Im } B \oplus \text{Ker } C$$

and that the zero pencil then becomes

$$(sI_{n-m} - NAM)$$

There are then (n-m) finite zeros, given by $\sigma(NAM)$, and there are m first-order asymptotic branches of the generalized Root-Locus and Nyquist diagrams given by:

$$s_{i,1} \simeq -\sigma_i(CB)/g \qquad g \to 0 \qquad i = 1,2,\ldots,m$$

and

$$g_{i,1} \simeq - \frac{\sigma_i(CB)}{j\omega} \qquad \omega \to \infty$$

where $\{\sigma_i(CB) : i = 1,2,...,m\}$ are the eigenvalues of the first Markov parameter CB.

The numbers of the various types of asymptotic behaviour are summarized in the following table (where ν is the first integer for which the nullity d_ν vanishes):

Order	Number of Nyquist Asymptotes	Number of Root Locus Asymptotes
1	$m - d_1$	$m - d_1$
2	$d_1 - d_2$	$2(d_1 - d_2)$
3	$d_2 - d_3$	$3(d_2 - d_3)$
⋮		
ν	$d_{\nu-1}$	$\nu d_{\nu-1}$
Total Number of Asymptotes	m	$m + \sum_{i=1}^{\nu-1} d_i$

Adding up the total number of root-locus asymptotes and subtracting the result from the dynamical order of the feedback loop we find that

$$\text{Number of finite zeros} = n - m - \sum_{i=1}^{\nu-1} d_i$$

A careful study of the implications of the above relationships gives a vivid geometrical insight into the role played by the basic state-space model operators A, B and C in generating the main structural features of the asymptotic behaviour of generalized Root-Locus and Nyquist diagrams.

2.6 Angles of Arrival at Zeros and Angles of Departure From Poles

The ways in which branches of the root locus depart from

poles and arrive at zeros, together with the relationships of other important structural features of these diagrams to the basic state-space model parameters has been considered by a number of investigators[7,9,33,34]. In this section we follow the approach of Thompson, Stein and Laub[34]. For the output-feedback system

$$\dot{x} = Ax + Bu$$
$$y = Cx$$
$$u = -\frac{1}{g} Ky$$

the closed-loop frequency matrix will be

$$S_K(g) = A - \frac{1}{g} BKC$$

Let s_i, x_i and y_i^* be closed-loop characteristic frequencies with their associated right and left eigenvectors so that for $0 < g < \infty$

$$\begin{aligned} [S_K(g) - s_i I]x_i &= 0 \\ y_i^*[S_K(g) - s_i I] &= 0 \end{aligned} \qquad i = 1,2,\ldots,n$$

then s_i, x_i, and y_i^* can be obtained by solving the generalized eigenvalue problems specified by

$$\begin{pmatrix} A-s_iI & B \\ -C & -gK^{-1} \end{pmatrix} \begin{pmatrix} x_i \\ w_i \end{pmatrix} = 0$$

$$i = 1,2,\ldots,p$$

$$[y_i^* \eta_i^*] \begin{pmatrix} A-s_iI & B \\ -C & -gK^{-1} \end{pmatrix} = 0$$

where role of integer p is discussed below. Then, in terms of these quantities it has been shown by Thompson et al[34] that:

The angles of the root locus for $0 \leq g \leq \infty$ and for distinct frequencies s_i are given by

$$\arg(ds_i) \;=\; \arg \frac{-y_i^* BKC\, x_i}{y_i^* \, x_i} \qquad 0 < g \leq \infty \qquad (2.10)$$

$$i = 1,2,\ldots,p$$

or

$$\arg(ds_i) \;=\; \arg \frac{\eta_i^* K^{-1} w_i}{y_i^* \, x_i} \qquad 0 \leq g < \infty \qquad (2.11)$$

$$i = 1,2,\ldots,p$$

The angles of departure from poles are found using (2.10) with $g = \infty$ and the angles of arrival at zeros are found using (2.11) with $g = 0$. Note that for $g > 0$ we will have $p = n$ and for $g = 0$ we will have $0 \leq p \leq n - m$, where p is the number of finite zeros and m is the number of inputs and outputs of the vector feedback loop.

2.7 Properties of Nyquist and Root Locus Diagrams For Optimal Feedback Systems

Several investigators have studied the asymptotic Nyquist diagram behaviour (for high frequencies), the asymptotic Root-Locus diagram behaviour (as explained below) and the location of the finite zeros for the standard optimal state feedback control problem involved in minimizing the cost function

$$J \;=\; \int_0^\infty (x^T Q x + \rho u^T R u)\,dt \qquad (2.12)$$

where R is positive definite symmetric, Q is positive semi-definite symmetric and ρ a finite positive real constant[10,34,35,36,37,38,51].

It is known as a result of these studies that:

(i) All the finite zeros of the optimal feedback loop lie in the left-half complex plane.

(ii) All the generalized Nyquist diagrams for the optimal feedback loop have infinite gain margin at least 60° phase margin.

(iii) When ρ in the performance index (2.12) is ∞ the n branches of the optimal root locus (i.e. the locus of closed-loop characteristic frequencies traced out with variation of ρ) start on a set of poles which are the stable poles of the original plant together with the mirror images of the imaginary axis of the unstable poles of the original plant (assumed to have no purely imaginary poles). As ρ varies all the branches of the optimal root locus remain in the left-half of the frequency plane. For ρ tending to zero a number p of the branches (where $0 \leq p \leq n-m$) stay finite and approach a set of loop transmission zeros. The remaining (m-p) branches approach infinity in a set of so-called Butterworth patterns. A kth order Butterworth pattern has k asymptotes each of which radiates from the origin through the left-half plane solutions of

$$s^{2k} = (-1)^{k+1}$$

2.8 Design Techniques

Despite the great efforts which have been expended on the problems involved over the past decade, much remains to be done in formulating a definitive design technique for linear multivariable feedback systems of high (say having up to 100 state variables) dynamic order. A fairly comprehensive attempt to extend the philosophy and techniques of the classical Nyquist-Bode-Evans design approach to the multivariable case is given in ref. 12, and this line of attack can be compared with a variety of other approaches to a common design problem in ref. 39. A fairly complete review of the development of the classical frequency-response methods can be found in ref. 13, together with a discussion of various forms of extension to the multivariable case. A computer-aided interactive-design package[40] has been developed for use in the implementation of the techniques described in ref. 12. Experience of its use has shown that an interactive design method of this sort can be useful to industrial designers. There is thus a considerable incentive to develop further those forms of design technique which are based on a combination of the attractive graphical features of complex-variable methods and the geometrical way in which the main structural features of the algebraic curves used are related to state-space model parameters.

Two ways in which extensions of classical frequency-response approaches to the multivariable case are being developed currently can be summarised as follows.

(i) The characteristic gain (Nyquist) loci and/or the characteristic frequency (Root-Locus) loci and their associated characteristic vector structures are manipulated using a set of ad-hoc rules (guided of course by engineering insight and using some computer-aided synthesis methods) until a set of engineering design criteria for relative stability and closed loop performance are satisfied. This approach has been described in ref. 12 and has been successfully deployed on a number of problems[12,39]. Although it requires engineering insight and manipulative skill, it has been shown to be a powerful tool in the hands of a skilled designer. A fairly extensive computer-aided design package has been developed for use in this way[40,41].

(ii) An alternative approach has been developed by Edmunds[42] which transfers the burden of the designer's choice to the selection of a closed-loop performance specification and a controller structure. This approach uses Nyquist and Bode diagram arrays of the open-loop and closed-loop behaviour as a means of assessing performance and interaction. Following a preliminary analysis of the problems (using the techniques of (i) above) the designer chooses a suitable specification for the closed-loop behaviour of the feedback system together with a stipulated structure for the feedback controller to be used. The controller structure selected has a number of variable parameters and the design is completed by a computer optimization (using say a standard weighted least squares algorithm) of the difference between the specified and actual closed-loop performance. As one would expect, this method, when properly used, produces better results than (i) above. It has the further advantage of being extendable to deal with variations in the system model parameters. The CAD package mentioned above[40,41] incorporates the routines for dealing with this approach. However, much further work remains to be done for a variety of reasons, including the following ones.

(i) Generalized Nyquist and generalized Root-Locus diagrams give only a partial picture of feedback system behaviour - they are primarily concerned with the relative stability of the closed-loop system.

(ii) Although the picture of the closed-loop system behaviour which is given by the use of characteristic gains can be extended by the use of an appropriate set of characteristic directions (eigenvectors of transfer-function matrices like $G(s)$ for example), this still does not give an adequate description of system closed-loop performance because, as already noted, characteristic values (eigenvalues) do not fully describe the gain behaviour of an operator.

It seems that for multivariable feedback systems it is necessary to work simultaneously with two forms of operator decomposition to get an adequate description for design purposes - a characteristic (eigenvalue) decomposition for stability assessment and a principal (singular) value decomposition for gain assessment.

(iii) The design specification used must be extended to include the effects of plant uncertainty, noisy disturbances and noisy sensors.

(iv) The precise roles played by feedback, feedforward and prefiltering must be defined in the context of a prescriptive design procedure, and these roles must be coherently related to an appropriate interpretation of the overall performance specification.

2.9 Conclusions

In the wider context of the multivariable feedback stability problem, Nyquist diagrams and Root-Locus diagrams are seen to be algebraic curves derived from appropriate matrix-valued functions of a complex variable. From this point of view, the stability and design approaches of classical feedback theory can be thought of as simple exploitations of the theory of algebraic curves. It is thus possible, and perhaps even likely, that there will be considerable further developments of complex-function-based approaches to linear feedback systems of increasing dynamical complication which exploit appropriate properties of algebraic curves. Such an approach would have considerable intuitive appeal, and would be well suited for use with the increasingly powerful interactive-graphic terminals which are becoming available for use by design engineers. The utility and acceptability of such a development would be greatly enhanced if it were closely linked to the state-space models on which so much of current control theory is based. In the forms of generalization presented here, there is clear evidence of useful and intuitively-appealing links between the algebraic curve theory and the parameters of the state-space model; this is a most hopeful augury for the future.

References

1. Postlethwaite, I. and MacFarlane, A.G.J. A Complex Variable Approach to the Analysis of Linear Multivariable Feedback Systems, Lecture Notes in Control and Information Sciences, Vol. 12, Springer-Verlag, Berlin, 1979.

2. MacFarlane, A.G.J. and Postlethwaite, I. "The generalized Nyquist Stability Criterion and Multivariable Root Loci", International Journal of Control, Vol. 25, 1977 pp.581-622.

3. Barman, J.F. and Katzenelson, J. "A Generalized Nyquist type Stability Criterion for Multivariable Feedback Systems", International Journal of Control, Vol. 20, 1974, pp. 593-622.

4. Bohn, E.V. and Kasvand, I. "Use of Matrix Transformations and System Eigenvalues in the Design of Linear Multivariable Control Systems", Proceedings of the Institution of Electrical Engineers, Vol. 110, 1963, pp. 989-996.

5. MacFarlane, A.G.J. "The Return-difference and Return-ratio Matrices and their Use in the Analysis and Design of Linear Multivariable Feedback Control Systems", Proc. of the Inst. of Elec. Engrs., Vol. 117, 1970, pp. 2037-2049.

6. Retallack, D.G. "Extended Root-Locus Technique for Design of Linear Multivariable Feedback Systems", Proc. Instn. of Elec. Engrs., Vol. 117, 1970, pp.618-622.

7. Kouvaritakis, B. and Shaked, U. "Asymptotic Behaviour of Root Loci of Linear Multivariable Systems", Internl. J. of Control, Vol. 23, 1976, pp. 297-340.

8. MacFarlane, A.G.J., Kouvaritakis, B. and Edmunds, J.M. "Complex Variable Methods for Multivariable Feedback Systems Analysis and Design", Proc. of Symp. on Alternatives for Linear Multivariable Control, National Engineering Consortium, Chicago, 1978, pp. 189-228.

9. Kouvaritakis, B. and Edmunds, J.M. The Characteristic Frequency and Characteristic Gain Design Method for Multivariable Feedback Systems", Proc. of Symp. on Alternatives for Linear Multivariable Control, National Engineering Consortium, Chicago, 1978, pp. 229-246.

10. Kouvaritakis, B. "The Optimal Root Loci of Linear Multivariable Systems", Internl. J. of Control, Vol.28, 1978, pp. 33-62.

11. Owens, D.H. Feedback and Multivariable Systems, Peter Peregrinus, Hitchin, England, 1978.

12. MacFarlane, A.G.J., Ed., Complex Variable Methods for Linear Multivariable Feedback Systems, Taylor and Francis, London, 1980.

13. MacFarlane, A.G.J., Ed., Frequency Response Methods in Control Systems, IEEE Reprint Series, IEEE, New York, 1979.

14. MacFarlane, A.G.J. and Karcanias, N. "Relationships Between State-Space and Frequency-Response Concepts", Preprints of Seventh IFAC World Congress, Vol. 3, 1978, pp. 1771-1779.

15. Rosenbrock, H.H. State Space and Multivariable Theory, Nelson, London, 1970.

16. Moore, B.C. "Singular Value Analysis of Linear Systems, Parts I and II", Systems Control Reports 7801, 7802, July 1978, University of Toronto.

17. Klema, V.C. and Laub, A.J. "The Singular Value Decomposition : Its Computation and Some Applications", Laboratory for Information and Decision Systems Report LIDS-R-892, M.I.T., Cambridge MA. See also IEEE Trans. on Auto. Control, Vol. 25, April 1980.

18. Laub, A.J. "Linear Multivariable Control. Numerical Considerations", Laboratory for Information and Decision Systems, Report ESL-P-833, July 1978, M.I.T., Cambridge, MA.

19. MacFarlane, A.G.J. and Scott-Jones, D.F.A. "Vector Gain" International Journal of Control, Vol. 29, 1979,pp.65-91.

20. Postlethwaite, I., Edmunds, J.M. and MacFarlane, A.G.J. "Principal Gains and Principal Phases in the Analysis of Linear Multivariable Feedback Systems", Cambridge

University Engineering Dept., Report CUED/F-CAMS/TR201, 1980, Cambridge, England.

21. MacFarlane, A.G.J. and Karcanias, N. "Poles and Zeros of Linear Multivariable Systems : A Survey of Algebraic Geometric and Complex-Variable Theory", Int. J. Control, Vol. 24, 1976, pp. 33-74.

22. Daniel, R.W. "Rank-deficient Feedback and Disturbance Rejection", Int. J. Control, Vol. 31, 1980, pp.547-554.

23. Desoer, C.A. and Schulman, J.D. "Zeros and Poles of Matrix Transfer Functions and Their Dynamical Interpretation", IEEE Trans. on Circuits and Systems, Vol. CAS-21, 1974, pp. 3-8.

24. Kouvaritakis, B. and MacFarlane, A.G.J. "Geometric Approach to Analysis and Synthesis of System Zeros Part I : Square Systems, and Part 2 : Non-square Systems", Int. J. Control, Vol. 23, 1976, pp. 149-166 and pp. 167-181.

25. Karcanias, N. and Kouvaritakis, B. "The output and Zeroing Problem and its relationship to the Invariant Zero Structure : A Matrix Pencil Approach", Int. J. Control, Vol. 30, 1979, pp. 395-415.

26. Sastry, S.S. and Desoer, C.A. "Asymototic Unbounded Root Loci by the Singular Value Decomposition", Electronics Research Laboratory, College of Engineering, Memorandum UCB/ERL M79/63 August 1979, University of California, Berkeley, CA.

27. Postlethwaite, I. "The Asymptotic Behaviour, the Angles of Departure and the Angles of Approach of the Characteristic Frequency Loci", Int. J. Control, Vol. 25, 1977, pp. 677-695.

28. Edmunds, J.M. "Calculation of Root Loci and Zeros Using a Bilinear Transformation", Colloquium on Numerical Algorithms in Control, Institution of Elec. Engrs., London, 1979.

29. Kouvaritakis, B. and Shaked, U. "Asymptotic Behaviour of Root Loci of Linear Multivariable Systems", Int. J. Control, Vol. 23, 1976, pp. 297-340.

30. Kouvaritakis, B. and Edmunds, J.M. "Multivariable Root Loci : A Unified Approach to Finite and Infinite Zeros", Int. J. Control, Vol. 29, 1979, pp. 393-428.

31. Owens, D.H. "Dynamic Transformations and the Calculation of Multivariable Root Loci", Int. J. Control, Vol. 28, 1978, pp. 333-343.

32. Owens, D.H. "A Note on Series Expansions for Multivariable Root Loci", Int. J. Control, Vol. 25, 1977, pp. 819-820.

33. Shaked, U. "The Angles of Departure and Approach of the Root Loci in Linear Multivariable Systems", Int. J. Control, Vol. 23, 1976, pp. 445-457.

34. Thompson, P.M., Stein, G. and Laub, A.J. "Analysis Techniques for Multivariable Root Loci", Laboratory for Information and Decision Systems, Report LIDS-P-965, M.I.T., Cambridge, MA.

35. Kwakernaak, H. "Asymptotic Root Loci of Multivariable Linear Optimal Regulators", IEEE Trans. on Automatic Control, Vol. AC-21, 1976, pp. 378-382.

36. Shaked, U. and Kouvaritakis, B. "The Zeros of Linear Optimal Control Systems and their Role in High Feedback Gain Stability Design", IEEE Trans. on Auto. Control, Vol. AC-22, 1977, pp. 597-599.

37. Kalman, R.E. "When is a Linear Control System Optimal?", Trans. of ASME Journal of Basic Eng. Series D, Vol. 86, 1964, pp. 51-60.

38. Safonov, M.G. and Athans, M. "Gain and Phase Margins for Multiloop LQG Regulators", IEEE Trans. Auto. Control, Vol. AC-22, 1977 pp. 173-179.

39. Sain, M., Peczkowski, J.L. and Melsa, J.I., Eds., Alternatives for Linear Multivariable Control, National Engineering Consortium, Chicago, 1978.

40. Edmunds, J.M. "Cambridge Linear Analysis and Design Programs", Cambridge University Engineering Department, Report CUED/F-CAMS/TR 198, 1980, Cambridge, England.

41. Edmunds, J.M. "Cambridge Linear Analysis and Design Programs", Proc. of IFAC Symp. on Computer-Aided Control System Design, Zurich, Switzerland, 1979.

42. Edmunds, J.M. "Control System Design and Analysis Using Closed-Loop Nyquist and Bode Arrays", Int. J. Control, Vol. 30, 1979, pp. 773-802.

43. Wonham, W.M. Linear Multivariable Control : A Geometric Approach, 2nd ed., Springer-Verlag, Berlin, 1979.

44. Wonham, W.M. "Geometric State-Space Theory in Linear Multivariable Control : A Status Report", Automatica, Vol. 15, 1979, pp. 5-13.

45. Moore, B.C. "On the Flexibility Offered by State Feedback in Multivariable Systems Beyond Closed-Loop Eigenvalue Assignment", IEEE Trans. Automatic Control, Vol. AC-21, 1976, pp. 689-691.

46. Harvey, C.A. and Stein, G. "Quadratic Weights for Asymptotic Regulator Properties", IEEE Trans. on Automatic Control, Vol. AC-23, 1978, pp. 378-387.

47. Stein, G. "Generalized Quadratic Weights for Asymptotic Regulator Properties", IEEE Trans. on Automatic Control Vol. AC-24, 1979, pp. 559-566.

48. Grimble, M.J. "Design of Optimal Output Regulators Using Multivariable Root Loci", Sheffield City Polytechnic, Report EEE/37, April 1979, Sheffield, England.

49. Safonov, M.G. "Choice of Quadratic Cost and Noise Matrices and the Feedback Properties of Multiloop LQG

Regulators", Proc. Asilomar Conf. on Circuits, Systems and Computers, Pacific Grove CA, November, 1979.

50. Doyle, J.C. "Robustness of Multiloop Linear Feedback Systems", Proceedings of 17th IEEE Conf. on Decision and Control, San Diego, CA, January, 1979.

51. Postlethwaite, I. "A Note on the Characteristic Frequency Loci of Multivariable Linear Optimal Regulators", IEEE Trans. Automatic Control, Vol. AC-23, 1978, 757-760.

52. Valenca, J.M.E. and Harris, C.J. "Nyquist criterion for input/output stability of multivariable systems", Int. J. Control, Vol. 31, 1980, 917-935.

53. Desoer, C.A. and Wang, Y.-T. "On the Generalized Nyquist Stability Criterion", IEEE Trans. Automatic Control, Vol. AC-25, 1980, 187-196.

Chapter 3

Robustness in multivariable control system design

Dr. IAN POSTLETHWAITE

Synopsis

In this lecture, techniques are described for assessing the relative stability of a multivariable control system design. An interesting and important feature in multivariable feedback systems is that they have different stability margins at different points in their configuration.

3.1 Introduction

In the generalized Nyquist stability criterion put forward by MacFarlane[6], and subsequently proved by several researchers[1,2,7], the stability of a linear, time-invariant, *multivariable*, feedback system, such as is shown in Fig. 3.1, is determined from the *characteristic gain loci* which are a plot of the eigenvalues of the return-ratio matrix $G(s)K(s)$ (or equivalently the eigenvalues of the return-ratio matrix $K(s)G(s)$) as the Laplace transform variable s traverses the standard Nyquist D-contour. The criterion can be stated as follows:

Generalized Nyquist Stability Criterion. The feedback system is stable if, and only if, the number of anticlockwise encirclements of the critical point -1 by the characteristic gain loci, is equal to the number of open-loop unstable poles.

If a gain parameter k is introduced into each of the loops then closed-loop stability can be checked on the same plot for a family of parameter values k, by counting

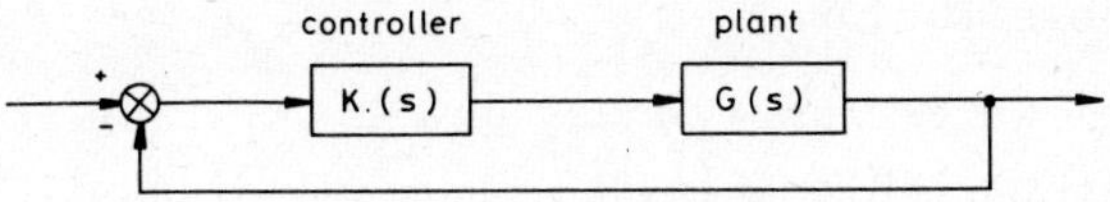

Fig. 3.1 Feedback system.

encirclements of $-1/k$. Also, in analogy with the scalar situation, gain and phase margins can be defined from the characteristic gain loci which indicate the limiting values of the modulus and phase that k can attain before instability occurs. Although these features are very attractive the characteristic gain loci do not by themselves give reliable information concerning the robustness of the closed-loop stability property. For example, the characteristic gain loci can have infinite gain margins with respect to the gain parameter k, and only a small change in some other system parameter may make the closed-loop system unstable.

In search of a reliable means of assessing the robustness of the closed-loop stability property, Doyle and Stein[3] introduced a gain margin concept defined in terms of the maximum singular value of an appropriate frequency-response matrix. They showed that the closed-loop system remains stable when subjected to a stable matrix perturbation $I+P(s)$, at any point in the configuration, providing the maximum singular value of $P(j\omega)$ multiplied by the maximum singular value of the frequency-response matrix $[I+T(j\omega)^{-1}]^{-1}$ is less than 1, at all frequencies, where $T(s)$ is the return-ratio matrix for the point in question. Note that since the return ratio matrix is generally different at different points in the configuration it follows that the robustness of the stability property will also vary around the system, and in particular it will vary before and after the plant where uncertainties are most likely to occur. Following the work of Doyle and Stein[3], Postlethwaite, Edmunds and MacFarlane[8] introduced a multivariable phase margin which in conjunction with Doyle and Stein's gain margin, allows a larger class of perturbations to be identified for which the feedback system remains stable. These results are discussed in the latter half of this lecture.

A system which has a large degree of stability in terms of these multivariable gain and phase margins will have characteristic gain loci which although not necessarily insensitive to small perturbations will nevertheless be insensitive enough for the number of encirclements of the critical point to remain unchanged. In the first half of this lecture a direct study is made of the sensitivity of the characteristic gain loci to small perturbations. *Sensitivity indices* are defined in terms of the left-hand and right-hand eigenvectors of the return-ratio matrix corresponding to the point at which the system is perturbed. It follows that the characteristic gain loci are at their least sensitive when the corresponding return-ratio matrix is normal.

3.2 Sensitivity of the Characteristic Gain Loci

In this section the perturbations of the eigenvalues of a return-ratio matrix $T(j\omega)$ are considered as the matrix is perturbed to $T(j\omega) + \varepsilon E(j\omega)$, where ε is a small, positive, real number. The analysis is based on two theorems due to Gerschgorin[5], and follows exactly the treatment given by Wilkinson[11] in his excellent monograph on the numerical aspects of the algebraic eigenvalue problem. Because of this and in order to concentrate on the main ideas, some details and proofs will be omitted; the interested reader is strongly encouraged to study Wilkinson's book.

For convenience, the $j\omega$'s are sometimes omitted, the notation t_{ij} will be used for the ijth element of the $m \times m$ matrix T, and $|t_{ij}|$ will denote the modulus of t_{ij}.

Theorem 1 (Gerschgorin)[5]

Each eigenvalue of T lies in at least one of the discs

with centres t_{ii} and radii $\sum_{i \neq j} |t_{ij}|$.

<u>Theorem 2</u> (Gerschgorin)[5]

If n of the circular discs of Theorem 1 form a connected domain which is isolated from the other discs, then there are precisely n eigenvalues of T within this connected domain.

3.2.1 Sensitivity indices

In the analysis that follows a set of quantities will be used repeatedly. They are of major importance and so will be introduced first. They are defined by the relations

$$s_i = \underline{y}_i^T \underline{x}_i \qquad i = 1,2,\dots,m$$

where $\underline{y}_i^T$ is $\underline{y}_i$ transposed, and $\underline{y}_i, \underline{x}_i$ are (respectively) left-hand and right-hand eigenvectors of T, normalized so that the sum of the squares of the moduli of their elements is equal to unity.

As will be seen later the m numbers $|s_i|$ give a direct indication of the sensitivity of the m eigenvalues of T and will therefore be referred to as the *sensitivity indices* of T. It is clear that the sensitivity indices range between 0 and 1 .

3.2.2 Analysis

In the generic situation where T has linear elementary divisors there exists a non-singular matrix W such that

$$W^{-1}TW = \text{diag}(\lambda_i)$$

where the λ_i are the eigenvalues of T, and diag(λ_i) is a diagonal matrix of the λ_i. It is immediately apparent that the columns of W are a candidate set of right-hand eigenvectors for T, and also that the rows of W^{-1} are a candidate set of left-hand eigenvectors for T. Further, if the column vectors of W are normalized to unity then the ith column of W can be taken to be $\underline{x}_i$, and the ith row of W^{-1}

taken to be $\underline{y}_i^T/s_i$.

With this definition for W it follows that

$$W^{-1}(T+\varepsilon E)W = \mathrm{diag}(\lambda_i) + \varepsilon \begin{pmatrix} e_{11}/s_1 & e_{12}/s_1 & \cdots & e_{1m}/s_1 \\ e_{21}/s_2 & e_{22}/s_2 & \cdots & e_{2m}/s_2 \\ \cdot & \cdot & \cdot & \cdot \\ e_{m1}/s_m & e_{m2}/s_m & \cdots & e_{mm}/s_m \end{pmatrix}$$

and the application of Theorem 1 to this shows that the perturbed eigenvalues lie in discs with centres $(\lambda_i+\varepsilon e_{ii}/s_i)$ and radii $\varepsilon \sum_{j\neq i} |e_{ij}/s_i|$. If all the eigenvalues are simple then for a sufficiently small ε all the discs are isolated and by Theorem 2 each disc contains precisely one eigenvalue. Alternatively, if there are some multiple eigenvalues, for example $\lambda_1 = \lambda_2 = \lambda_3$ and $\lambda_4 = \lambda_5$, then there will be three Gerschgorin discs with centres $\lambda_1 + \varepsilon e_{ii}/s_i$, $i = 1,2,3$ and two with centres $\lambda_4 + \varepsilon e_{ii}/s_i$, $i = 4,5$, all with radii of order ε. For sufficently small ε the group of three discs will be isolated from the group of two, but, in general, no claim of isolation can be made about the individual discs in a group.

From the above analysis it is clear that the sensitivity of the eigenvalues is directly related to the sensitivity indices; the larger s_i is, the less sensitive λ_i is to small perturbations in T.

It is now demonstrated, by example, how, as T approaches a matrix having non-linear elementary divisors, the corresponding eigenvalues become increasingly sensitive. To facilitate this let T be given by

$$T = \begin{pmatrix} t_{11} & 1 \\ 0 & t_{22} \end{pmatrix}$$

which approaches a simple Jordan submatrix as t_{11} approaches t_{22}. The eigenvalues of T are t_{11} and t_{22}, and the corresponding normalized right-hand and left-hand eigenvectors are given by

$$\underline{x}_1 = \begin{pmatrix} 1 \\ 0 \end{pmatrix}, \quad \underline{x}_2 = \frac{1}{\sqrt{[1+(t_{11}-t_{22})^2]}} \begin{pmatrix} 1 \\ t_{22}-t_{11} \end{pmatrix}$$

$$\underline{y}_1 = \frac{1}{\sqrt{[1+(t_{11}-t_{22})^2]}} \begin{pmatrix} t_{11}-t_{22} \\ 1 \end{pmatrix}, \quad \underline{y}_2 = \begin{pmatrix} 0 \\ 1 \end{pmatrix}$$

where for convenience t_{11} and t_{22} have been assumed real. The sensitivity indices are therefore given by

$$|s_1| = |\underline{y}_1^T \underline{x}_1| = |(t_{11}-t_{22})/\sqrt{[1+(t_{11}-t_{22})^2]}|$$

$$|s_2| = |\underline{y}_2^T \underline{x}_2| = |(t_{22}-t_{11})/\sqrt{[1+(t_{11}-t_{22})^2]}|$$

and it is seen that both indices tend to zero as t_{11} approaches t_{22}. Hence from the previous analysis both simple eigenvalues become increasingly sensitive as T tends to the simple Jordan submatrix.

3.3 Uncertainty in a Feedback System

This section considers the implications of the sensitivity results of Section 3.2 to the design of feedback systems.

For the feedback configuration of Fig. 3.1 the characteristic gain loci are the eigenvalues of the return-ratio matrix $G(j\omega)K(j\omega)$, or alternatively the eigenvalues of the return-ratio matrix $K(j\omega)G(j\omega)$. Although the eigenvalues of the two return-ratio matrices are the same, the corresponding eigenvectors will generally be different, and hence the characteristic gain loci may be sensitive to perturbations in $G(j\omega)K(j\omega)$ while insensitive to perturbations in $K(j\omega)G(j\omega)$. To see the significance of this consider the feedback configuration shown in Fig. 3.2, where the return-ratio matrix $G(j\omega)K(j\omega)$ is assumed to have been perturbed to $G(j\omega)K(j\omega) + \varepsilon E(j\omega)$. Insofar as stability is concerned, this configuration is equivalent to that shown in Fig. 3.3, where

$$E_o(j\omega) = \varepsilon E(j\omega)[(G(j\omega)K(j\omega))]^{-1}$$

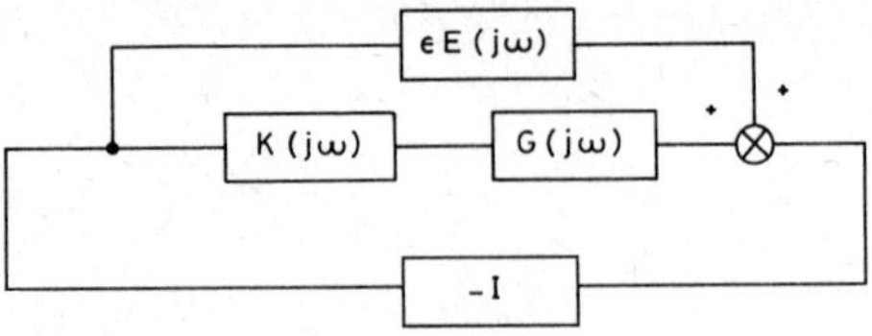

Fig. 3.2 Perturbation of return-ratio matrix $G(j\omega)K(j\omega)$

Therefore, the sensitivities of the eigenvalues of $G(j\omega)K(j\omega)$ give information concerning stability with respect to small perturbations (uncertainty) associated with the plant output. Note that at frequencies for which $[G(j\omega)K(j\omega)]^{-1}$ becomes very large, the moduli of the characteristic gain loci become very small, and hence their sensitivity becomes unimportant.

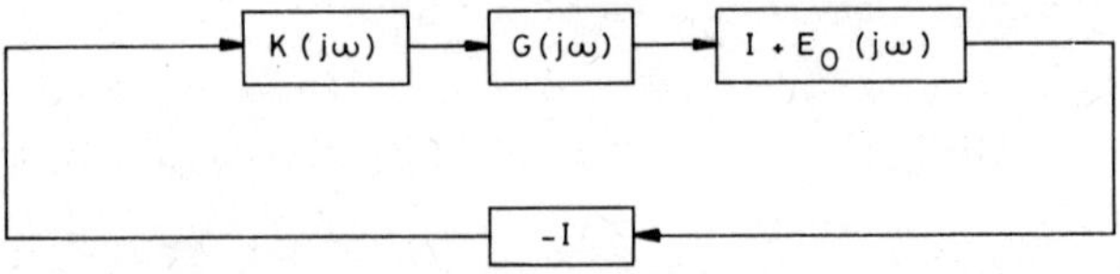

Fig. 3.3 Perturbation associated with plant output.

In an analogous fashion it can be shown that the sensitivities of the eigenvalues of $K(j\omega)G(j\omega)$ give information concerning stability with respect to small perturbations (uncertainty) associated with the plant input. For more complicated feedback configurations a return-ratio matrix can be defined at any particular point at which uncertainty is felt to be a problem and the sensitivity of the characteristic gain loci checked via the corresponding sensitivity indices. The return-ratio matrices defined for break points before and after the plant will normally be the most important, and will therefore be referred to as the *input and output return-ratio matrices*, denoted by $T_i(s)$ and $T_o(s)$ respectively.

3.3.1 Normality

The sensitivity indices are numbers between 0 and 1, and when they are all 1 it is clear that the characteristic gain loci are at their least sensitive. A class of matrices for which this is always the case is the set of normal matrices.

Diagonal matrices are normal and therefore design techniques which aim to decouple the outputs from the inputs using a precompensator are well founded from the point of view of achieving robustness with respect to small perturbations at the plant output. It does not follow, however, that such a design would necessarily be robust with respect to small perturbations at the plant input.

3.4 Relative Stability Matrices

In order to study the robustness of the closed-loop stability property with respect to large and possibly dynamic perturbations consider the feedback configuration of Fig. 3.4, where $I+\Delta G(s)$ represents a multiplicative perturbation of the plant from $G(s)$ to $G(s) + G(s)\Delta G(s)$. The configuration can be redrawn as shown in Fig. 3.5, where the perturbation $\Delta G(s)$ now appears in series with a transfer function $R_i(s)$, given by

$$R_i(s) \triangleq [I+(K(s)G(s))^{-1}]^{-1}$$

$R_i(s)$ will be called the *relative stability matrix* with respect to uncertainty at the plant input, or simply the *input relative stability matrix*.

For a single-input, single-output system a perturbation $1+\Delta g(s)$ commutes around the feedback loop and the relative stability function $[1+(k(s)g(s))^{-1}]^{-1}$ is the same at any point in the configuration. However, in multivariable systems this is not the case, and thus it becomes important to examine the effects of perturbations at different points in the loop. The most likely places for uncertainties to occur are at the plant input, considered above, and at the plant output. To analyse the effects of uncertainty at the output

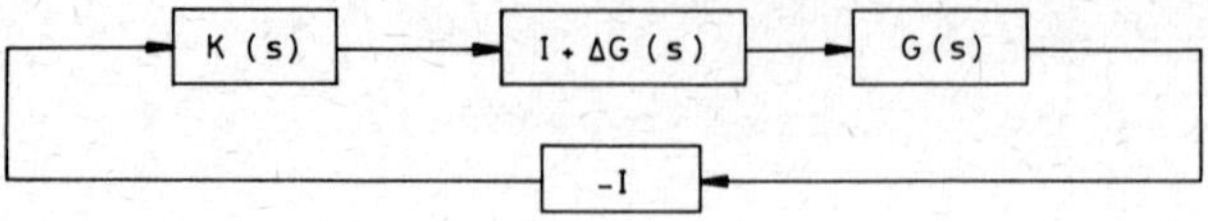

Fig. 3.4 Dynamic uncertainty at plant input.

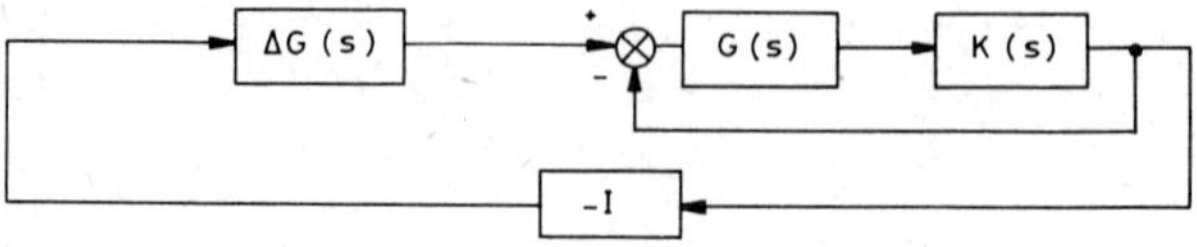

Fig. 3.5 A rearrangement of Fig. 3.4 .

consider a multiplicative perturbation, $I+\Delta G(s)$, after the plant shown in Fig. 3.6 . This configuration can be redrawn as shown in Fig. 3.7, where $\Delta G(s)$ now appears in series with a transfer function $R_o(s)$, given by

$$R_o(s) \triangleq [I+(G(s)K(s))^{-1}]^{-1}$$

$R_o(s)$ will be called the *relative stability matrix* with respect to uncertainty at the plant output, or simply the *output relative stability matrix*. It is interesting to note that the input and output relative stability matrices are related by the following expression:

$$R_i(s) = G(s)^{-1}R_o(s)G(s)$$

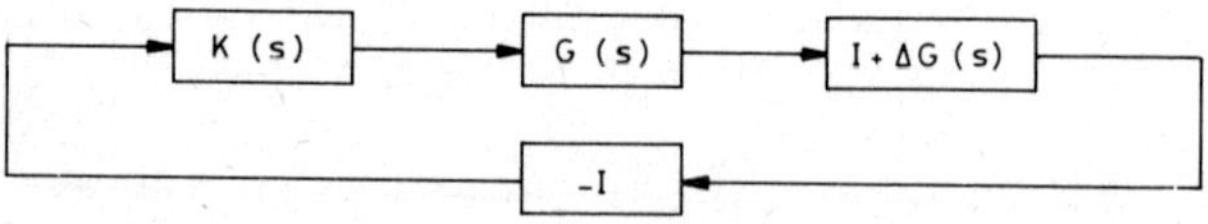

Fig. 3.6 Dynamic uncertainty at the plant output.

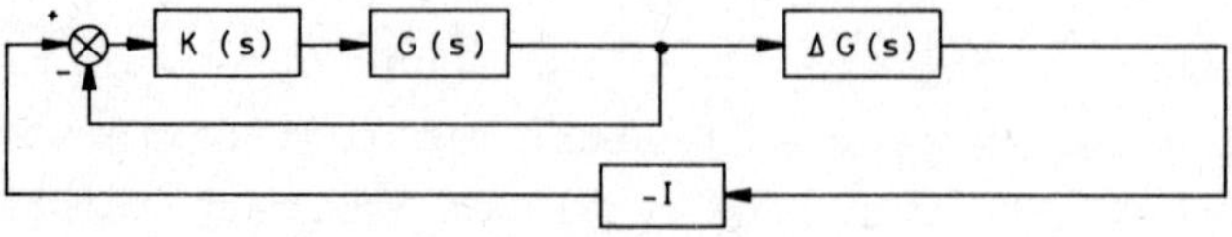

Fig. 3.7 A rearrangement of Fig. 3.6 .

In the next section, it is shown how multivariable gain and phase margins can be defined in terms of the relative stability matrices thereby characterizing perturbations for which the feedback system remains stable.

3.5 Multivariable Gain and Phase Margins

If $R(s)$ is the relative stability matrix corresponding to a multiplicative perturbation $I+\Delta G(s)$, at either the input or output of the plant, then conditions can be derived[8] for $\Delta G(s)$, in terms of $R(s)$, for which the perturbed closed-loop system remains stable. The conditions are summarized in theorems 5 and 6 after first introducing some definitions and notation.

The gain and phase information from which the multivariable gain and phase margins are derived stems from the polar decomposition of a complex matrix[4] which is now defined. Analogous to the polar form of a complex number, a complex matrix R can be represented in the forms

$$R = UH_R \tag{3.1}$$

$$R = H_L U \tag{3.2}$$

where U is unitary and H_R, H_L are positive semidefinite Hermitian matrices. H_R, H_L are often called right and left moduli, and are uniquely determined as $\sqrt{R^*R}$, $\sqrt{RR^*}$ respectively, where $*$ denotes complex conjugate transpose, and U is uniquely defined via (3.1) or (3.2) when R is nonsingular. Representations (3.1) and (3.2) are called *Polar Decompositions* of R, and are easily determined from the *Singular Value Decomposition* of R [10], for which there exists well tested software. If R has the singular value decomposition

$$R = X\Sigma V^* \; ; \quad X,V \text{ unitary, and } \Sigma \text{ diagonal and } \geq 0$$

then $$R = (XV^*)(V\Sigma V^*) = UH_R$$

and $$R = (X\Sigma X^*)(XV^*) = H_L U$$

From the polar decompositions the following key quantities are defined:

(i) The eigenvalues of R are called *characteristic gains.*

(ii) The eigenvalues of the Hermitian part in either polar decomposition of R are called *principal gains,* or *singular values.*

(iii) The arguments of the eigenvalues of the unitary part of the polar decompositions of R are called *principal phases.*

These three quantities are related by two important theorems.

Theorem 3 The magnitudes of the characteristic gains are bounded above and below by the maximum and minimum principal gains.

Theorem 4 If the principal phases have a spread of less than π, then the arguments of the characteristic gains are bounded above and below by the maximum and minimum principal phases.

Based on these theorems Postelthwaite, Edmunds and MacFarlane[8] derived the reobustness results discussed in the remainder of this section.

Let $R(j\omega)$ have principal gains and principal phases

$$\alpha_1(\omega) \leq \alpha_2(\omega) \leq \ldots \leq \alpha_m(\omega)$$

and

$$\theta_1(\omega) \leq \theta_2(\omega) \leq \ldots \leq \theta_m(\omega)$$

respectively; it is assumed that the $\{\theta_i(\omega)\}$ have a spread of less than Π. Similarly, let the principal gains and phases of $\Delta G(j\omega)$ be

$$\delta_1(\omega) \leq \delta_2(\omega) \leq \ldots \leq \delta_m(\omega)$$

and

$$\varepsilon_1(\omega) \leq \varepsilon_2(\omega) \leq \ldots \leq \varepsilon_m(\omega)$$

respectively. Also, let the condition numbers of $R(j\omega)$ and

$\Delta G(j\omega)$, using the ℓ_2-induced norm, be $c_1(\omega)$ and $c_2(\omega)$, respectively, so that

$$c_1(\omega) \triangleq \alpha_m(\omega)/\alpha_1(\omega)$$

and

$$c_2(\omega) \triangleq \delta_m(\omega)/\delta_1(\omega)$$

and also define

$$\psi_m(\omega) \triangleq \tan^{-1}\left\{\frac{[c_1(\omega)-1]c_2(\omega)}{1-[c_1(\omega)-1]c_2(\omega)}\right\}$$

which will be referred to as a phase modifier. Then the following theorems can be stated.

Theorem 5 (Small gain theorem) The perturbed closed-loop system remains stable, if

(i) $\Delta G(s)$ is stable, and
(ii) $\delta_m(\omega)\alpha_m(\omega) < 1$, for all ω.

Theorem 6 (Small phase theorem) The perturbed closed-loop system remains stable, if

(i) $\Delta G(s)$ is stable,
(ii) $\{\theta_i(\omega)+\varepsilon_j(\omega) : i,j = 1,\ldots,m\}$ have a spread of less than π, for all ω,
(iii) $[c_1(\omega)-1]c_2(\omega) < 1$, for all ω,
(iv) $\varepsilon_1(\omega)+\theta_1(\omega)-\psi_m(\omega) > -\pi$, for all ω, and
(v) $\varepsilon_m(\omega)+\theta_m(\omega)+\psi_m(\omega) < \pi$, for all ω.

Corollary 1. By symmetry theorem 6 can be restated with $c_1(\omega)$ and $c_2(\omega)$ interchanged.

Corollary 2. As a consequence of theorems 5 and 6, the perturbed closed-loop system remains stable, if, for some frequency ω_b,

(i) the conditions of theorem 6 are satisfied

in the frequency range $[0,\omega_b]$, and

(ii) the conditions of theorem 5 are satisfied in the frequency range $[\omega_b,\infty]$.

Remark 1. Theorem 5 is exactly the small gain theorem introduced by Zames[12] for a more general class of systems. It is also the robustness result used in the LQG-based designs of Doyle and Stein[3] and Safonov, Laub and Hartmann[9].

Remark 2. Although it might be argued that theorem 5 gives an adequate characterization of the robustness of the stability property, the combination of theorems 5 and 6 allows a larger set of perturbations to be identified for which the closed-loop system remains stable.

Remark 3. If in corollary 1, $\Delta G(s)$ is simply assumed to be a phase change in each of the feedback loops, then $\Delta G(j\omega) = \mathrm{diag}\{\exp(j\varepsilon_i)\}$ and $c_2(\omega) = 1$, so that condition (iii) is always satisfied, and the phase modification $\psi_m(\omega)$ is always zero.

Remark 4. Typically a perturbation $\Delta G(s)$ representing small parameter variations and unmodelled high frequency dynamics will, at low frequencies, have a condition number of approximately 1, and small principal gains. At high frequencies its maximum principal gain will increase drastically, and the minimum principal phase will exceed 180 degrees lag. Consequently, when predicting perturbations for which the closed-loop systems remain stable, a combination of theorems 5 and 6, as in corollary 2, is useful.

Remark 5. Theorem 5 indicates that, at frequencies for which the maximum principal gain of the perturbation is large, the maximum principal gain of $R(j\omega)$ should be designed to be small; this will normally be the case at high frequencies. At low frequencies, theorem 6 indicates that a perturbation with large gains can be tolerated providing $R(j\omega)$ is designed to have a condition number close to 1 and a small spread of principal phases. This situation will normally be feasible at low frequencies but at high frequencies when the phase lag inevitably exceeds 180 degrees the maximum principal gain of $R(j\omega)$ has necessarily to be made small.

From the preceding remarks it is clear that useful information concerning the robustness of the closed-loop stability property, is readily obtained from Bode plots of the principal gains and phases of the input and output relative stability matrices. Loosely speaking, in going from the scalar to multivariable case, analysis moves from studying a single Bode magnitude and phase plot to studying, for each point of uncertainty, a band of Bode plots (see Figs. 3.8 and 3.9), defined by the maximum and minimum principal gains and phases of the appropriate relative stability matrix. Also, as an indicator of the robustness of the stability property, gain and phase margins can be defined in terms of the maximum principal gain and the minimum principal phase (maximum phase lag) by considering them to be standard Bode magnitude and phase plots for a scalar system.

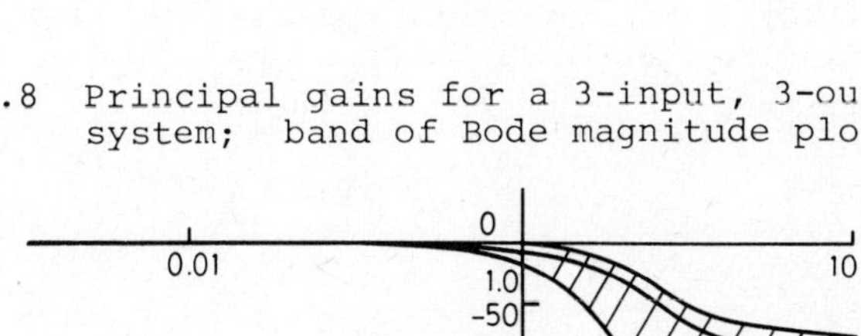

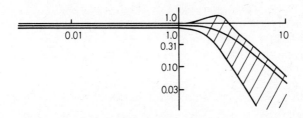

Fig. 3.8 Principal gains for a 3-input, 3-output system; band of Bode magnitude plots.

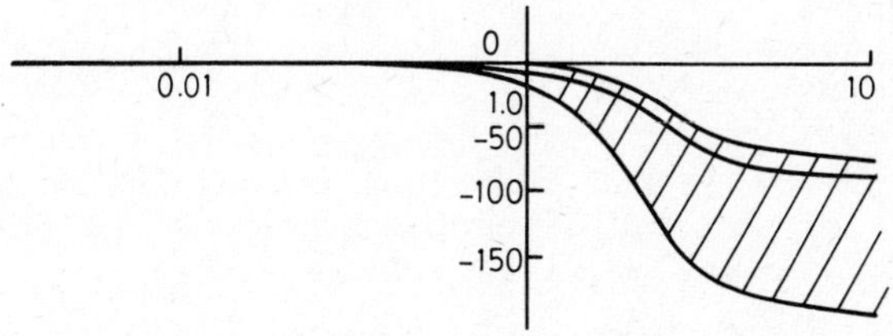

Fig. 3.9 Principal phases for a 3-input, 3-output system; band of Bode phase plots.

3.6 Conclusion

The *sensitivity indices* corresponding to the input and output return-ratio matrices indicate the sensitivity of the characteristic gain loci to small perturbations (uncertainties) at the input and output of the plant, respectively. The indices therefore, give crucial information concerning the possibility of closed-loop instability.

For large and possibly dynamic perturbations (uncertainties) the robustness of the closed-loop stability property is characterized by the *principal gains* and *principal phases* of the input and output relative stability matrices. From Bode-type plots of the principal values, multivariable gain and phase margins can be defined which are analogous to the classical single-loop stability margins.

REFERENCES

1. BARMAN, J.F. and J. KATZENELSON : A generalized Nyquist type stability criterion for multivariable feedback systems. Int. J. Control, 20, (1974), pp. 593-622.
2. DESOER, C.A. and Y.T. WANG : On the generalized Nyquist stability criterion. IEEE Trans. Autom. Control, 25, (1980), pp. 187-196.
3. DOYLE, J.C. and G. STEIN : Multivariable feedback design : concepts for a classical/modern synthesis. IEEE Trans. Autom. Control, 26, (1981), pp. 4-16.
4. GANTMACHER, F.R. : Theory of Matrices, Vol. 1, (1959) Chelsea, New York.
5. GERSCHGORIN, S. : Über die Abgrenzung der Eigenwerte einer Matrix. Iz.Akad.Nauk SSSR, (1931), 6, pp.749-754.
6. MACFARLANE, A.G.J. : Return-difference and return-ratio matrices and their use in analysis and design of multivariable feedback control systems. Proc. IEE, 117, (1970), pp. 2037-2049.
7. MACFARLANE, A.G.J. and I. POSTLETHWAITE : The generalized Nyquist stability criterion and multivariable root loci. Int. J. Control, 25, (1977), pp. 81-127.
8. POSTLETHWAITE, I., J.M. EDMUNDS and A.G.J. MACFARLANE : Principal gains and principal phases in the analysis of linear multivariable feedback systems. IEEE Trans. Autom Control, 26, (1981), pp. 32-46.
9. SAFONOV, M.G., A.J. LAUB and G.L. HARTMAN : Feedback properties of multivariable systems : the role and use of the return difference matrix. IEEE Trans. Autom. Control, 26, (1981), pp. 47-65.

10. STEWART, W.G. : Introduction to Matrix Computations. Academic, New York, (1973).

11. WILKINSON, J.H. : The Algebraic Eigenvalue Problem. Clarendon, Oxford, (1965).

12. ZAMES, G. : On the Input-Output Stability of Time Varying Nonlinear Feedback Systems. IEEE Trans. Autom. Control, 11, (1966), pp. 228-238.

PROBLEMS

P.3.1 In the feedback configuration of Fig. 3.1, the plant and controller are given by

$$G(s) = \frac{1}{(s+1)(s+2)} \begin{bmatrix} -47s+2 & 56s \\ -42s & 50s+2 \end{bmatrix}$$

$$K(s) = \begin{bmatrix} k_1 & 0 \\ 0 & k_2 \end{bmatrix}$$

where nominally $k_1 = k_2 = 1$.

The plant has been contrived so that it can be diagonalized at all frequencies using a constant matrix W. For the nominal controller parameters find W and hence the sensitivity indices, and plot the characteristic gain loci. Discuss the robustness of the closed-loop stability property with respect to variations in k_1 and k_2.

P.3.2 Figure 3.10 shows the principal gains and phases of the output relative stability matrix for an automotive gas turbine design study. Discuss briefly the relative stability of the design with respect to a dynamic perturbation $I+\Delta G(s)$ after the plant model.

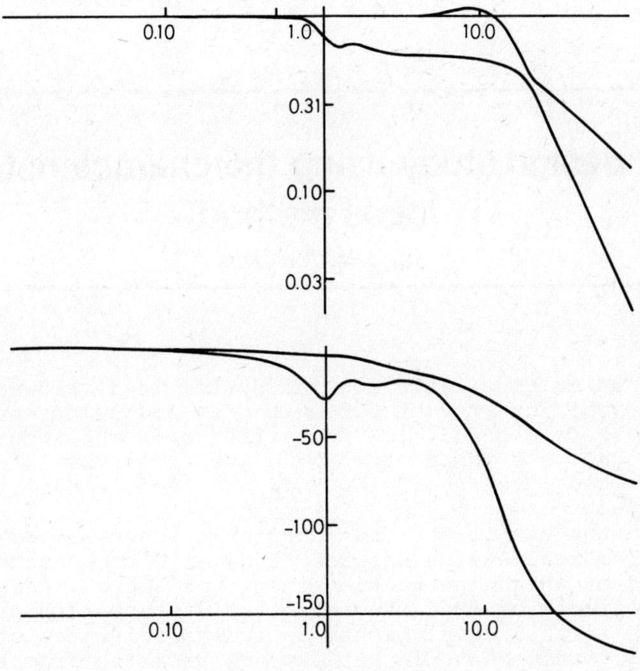

Fig. 3.10 Principal gains and phases of the output relative stability.

Chapter 4

A design study using the characteristic locus method

Dr. J. M. EDMUNDS

Synopsis

A 3-input multivariable control system is developed for the regulation of the product concentration and temperature of a 2-bed exothermic catalytic reactor. Two of the inputs are the flow rate and temperature of a quench stream injected between the beds; the third input is the feed temperature.

Through the use of the characteristic locus method of control system analysis, a cascade of dynamic compensators is developed by which interaction among variables is suppressed and the effects of feed concentration disturbances are controlled. In this development, an analysis is made of the potential improvements in performance accruing from the use of internal bed temperature measurements. The system also has the feature of resetting the quench flow rate to its nominal value upon sustained system disturbance.

4.1 Introduction

A multivariable control system for a two-bed catalytic reactor is synthesized here by a method that tailors the frequency-dependent eigen-properties of a transfer-function matrix to achieve control of both concentration and temperature of the reactor product stream. Such a control objective extends earlier investigations of single-output control, and the application of such a design technique complements earlier applications of the Linear-Quadratic-Gaussian technique to this reactor control problem[8,10]. The best possible control is sought, of course, with either design method, but here, instead of seeking an optimum of a quadratic criterion, the suppression of interaction between the two controlled variables is addressed directly.

To achieve control of both concentration and temperature

upon sustained disturbances in feed concentration, a control system configuration such as that shown in Figure 4.1 is proposed. An exothermic chemical reaction whose rate depends upon both temperature and concentration takes place in both beds of the reactor shown in this figure. It is this coupling of variables through the reaction rate that suggests the control configuration shown. Measurements of the controlled variables C_{out} and T_{out} are assumed available, and measurements of fluid temperatures at three points internal to each bed may also be considered available for use in

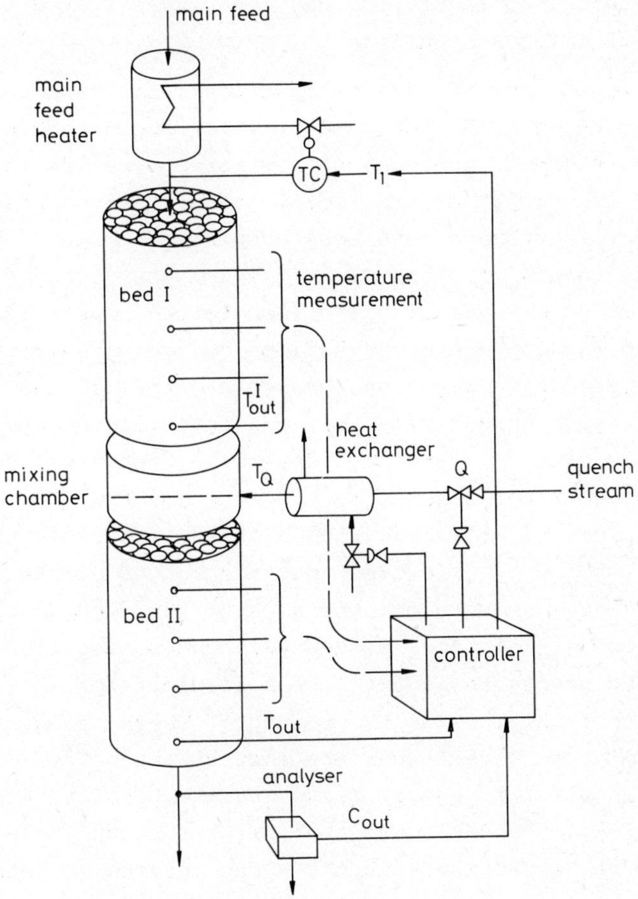

Fig. 4.1 Diagram of two-bed reactor and proposed control configuration for the regulation of C_{out} and T_{out}.

the control system should they prove beneficial. In industrial circumstances, measurement of C_{out} may be impossible, in which case a reconstruction of this variable from temperature measurements may be attempted as demonstrated by Silva and Wallman. Here, for purposes of system analysis, concentration is considered measurable. Three manipulatable inputs (main feed temperature T_f, quench temperature T_Q, and quench flow rate Q) are to be used. Such a control configuration is applicable in the frequently encountered industrial circumstance where conditions only at the entrance of each bed can be adjusted. With the exception of the input T_f, such a configuration has been investigated in the recent work of Silva and Wallman.

Following the results of that work, it is proposed to employ the quench flow rate only for short-term corrective action, forcing the quench flow rate to return to its nominal value as quickly as practicable after the transient effects of the disturbance have been suppressed. Such a stipulation leaves inputs T_f and T_Q to eliminate steady state offsets in C_{out} and T_{out}. Temperature inputs are best suited for long-term corrections owing to their relatively slow rate of propagation through the reactor and their relatively high sensitivity. It can be argued on physical grounds that T_f and T_Q determine steady-state concentration and temperature at the entrance of Bed II and thus are sufficient to eliminate steady-state offsets in the variables C_{out} and T_{out} provided that the ranges of the manipulated inputs are adequate for the size of the feed disturbance. However, it is not so evident without some analysis that this 3-input configuration will be sufficient for the control of dynamic conditions in the reactor. There are questions of the relative effect that the inputs have on the outputs over the important frequency range and the benefits accruing from the creation and placement of zeros through the use of temperature measurements at the three internal points in each bed.

The synthesis of dynamic compensators for this control system configuration is carried out here in two stages. First, because Bed II is obviously the important dynamic element of this configuration, a 2-input 2-output control system for this bed is synthesized and analyzed. Bed I and the associated third input T_f is then added serially to controlled Bed II with the additional requirement that the quench flow rate is to be driven by integral action to its nominal value at long times. All of these design stages are carried out with the use of locally linear 7th-order state-space models of the reactor beds of the form

$$\dot{x} = Ax + Bu$$

$$y = Cx + Du$$

The approach to the design, however, was not rigidly proscribed by these models but was influenced by the knowledge that high-order process modes were lacking in the models. To investigate the robustness of the designs, simulations of closed-loop performance were made using 14th-order linear models of the reactor, thus introducing the potentially troubling phase lag of unmodelled modes. The design technique used is the characteristic locus method[5]. The calculations were conveniently carried out on a minicomputer with an interactive program specially developed for linear system analysis[1].

4.2 Two-Bed Reactor Model

The reactor model used in these design calculations derives from a set of nonlinear partial differential equations representing one-dimensional material and heat balances of the packed bed. The equations and methods of their reduction to linear state-space form have been detailed by Silva et al[8]. and Michelson et al[6], and are not repeated here. Rather it is sufficient for purposes of this study to state the A,B,C and D matrices for each bed and the coupling between the beds.

A sketch showing the input-output configuration of the 2-bed system is given in Figure 4.2 . The models associated with each block of this figure are as follows.

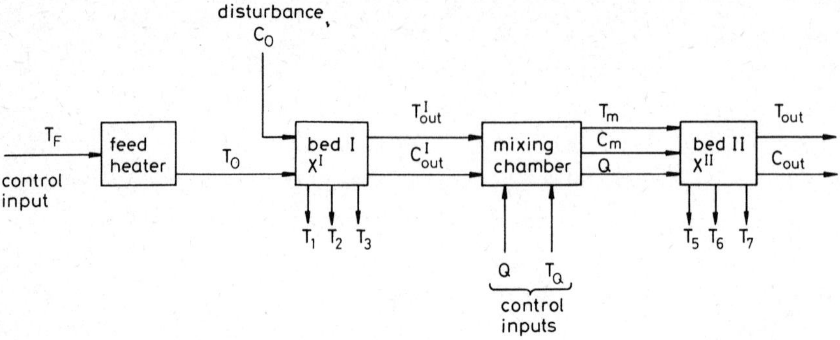

Fig. 4.2 Reactor system components and input-output variables. Notation: T = temperature, C = concentration, Q = flow rate, X = state vector.

Feed heater

$$\dot{T}_o = -\frac{1}{\tau} T_o + \frac{1}{\tau} T_F$$

Bed 1

$$\dot{X}^I = A^I X^I + B^I \begin{pmatrix} T_o \\ C_o \end{pmatrix}$$

$$\begin{pmatrix} T_1 \\ T_2 \\ T_3 \\ T^I_{out} \\ C^I_{out} \end{pmatrix} = C^I X^I + D^I \begin{pmatrix} T_o \\ C_o \end{pmatrix}$$

Mixing chamber

$$\begin{pmatrix} T_m \\ C_m \\ Q \end{pmatrix} = M \begin{pmatrix} Q \\ T_Q \\ T^I_{out} \\ C^I_{out} \end{pmatrix}$$

Bed 11

$$\dot{X}^{II} = A^{II} X^{II} + B^{II} \begin{pmatrix} T_m \\ C_m \\ Q \end{pmatrix}$$

$$\begin{bmatrix} T_5 \\ T_6 \\ T_7 \\ T_{out} \\ C_{out} \end{bmatrix} = C^{II}X^{II} + D^{II} \begin{bmatrix} T_m \\ C_m \\ Q \end{bmatrix}$$

The relations for the mixing chamber are linearizations of the steady-state material and heat balances at that point.

Reactor models of two different orders are used here. The design is carried out using a 7th-order model for each bed. These models were obtained from 14th-order models by a reduction technique in which the fast modes are discarded but the steady state gains are preserved[9]; catalyst temperatures were retained as the state variables. The 14th-order model is used here in closed-loop simulations to test the performance of the control system designed on the basis of the lower order model. The 14th-order model was not used for the design calculations since the software available at the time the design was being carried out could not handle all 28 states.

Elements of the matrices of the 7th-order reactor model and the mixing matrix are given in the appendix. The numerical values are identical to those determined for the experimental conditions of Silva and represent reactor dynamic behaviour in the vicinity of the nominal steady state of those experiments. Matrices for the 14th-order model are given by Silva[7].

4.3 Control System Design for Bed II

The control system for Bed II will employ the inputs Q and T_Q shown in Figure 4.2 to regulate effluent variables T_{out} and C_{out}. One of the first matters to be investigated in such a multivariable system is the sensitivity of the outputs to the inputs. To achieve independent control over T_{out} and C_{out}, one would prefer the transfer function matrix G(s) relating outputs to inputs to be well conditioned over all frequencies. Regulation of T_{out} and C_{out} to the setpoint, for example, requires that G(s=0) to be invertible. The situation for this reactor transfer function matrix is reflected in the open-loop Nyquist array, the characteristic loci, and the eigenvectors of G(s).

The first two of these are displayed in Figures 4.3 and 4.4 and reveal that we begin the design under inauspicious circumstances. The close similarity of the columns of the Nyquist array (Figure 4.3) and the large disparity of the eigenvalues of G(s) (Figure 4.4) would both seem to imply a near linear dependence of the inputs Q and T_Q for this set of outputs. However, the determinant of the steady state gain matrix G(0) is only 10 to 20 times smaller than the individual elements and thus suggests that control to setpoint is feasible. A further indicator of the feasibility of this 2x2 control configuration for Bed II is the angle between the two eigenvectors of G(2). Significant deviation from orthogonality of the eigenvectors implies

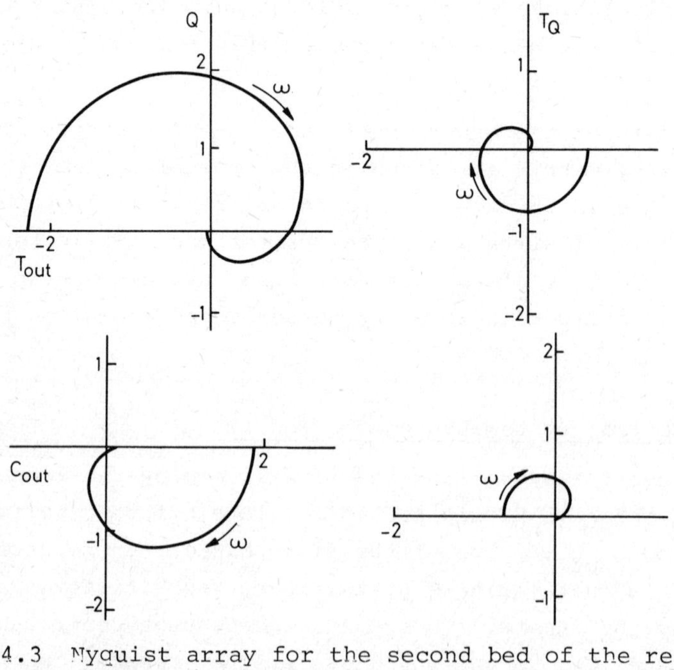

Fig. 4.3 Nyquist array for the second bed of the reactor. Outputs (T_{out}, C_{out}); Inputs (Q, T_Q).

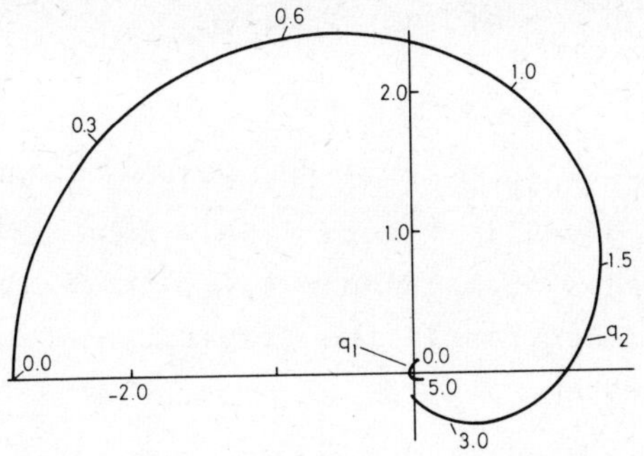

Fig. 4.4 Characteristic loci of Bed II.

the need for inordinately large control inputs. The angle between the two vectors for Bed II was found to be about 70° in the range of frequencies 0.1 to 3, the range above which model accuracy cannot be assured. The situation seems to be favourable; angles at the intermediate and high frequencies appear significantly large, and it is these that are relevant because Q is not used for low frequency corrective action.

4.4 Compensator Design

The addition of compensators for purposes of forging the characteristic loci into desired shapes proceeds through four distinguishable stages: (1) suppression of interaction at high frequency, (2) attenuation of the loci in the very high frequency range for which the model is poor, (3) boosting of gain at intermediate frequencies, and (4) addition of the integral action to enhance steady state accuracy and to suppress low frequency interaction.

Suppression of interaction at high frequencies is accomplished as described above by introducing a constant precompensator K_h that approximately aligns the eigenvectors of $G(s)K_h$ with the unit vectors at a selected frequency, which should be approximately the desired cut-off frequency.

The K_h matrix obtained for the alignment at $\omega = 2$ is

$$K_h = \begin{bmatrix} -1.16 & 1.04 \\ -3.35 & 2.34 \end{bmatrix} \quad (4.1)$$

and the characteristic loci of the reactor with this pre-compensator are given in Figure 4.5 . Upon comparison with the uncompensated system (Figure 4.4), there is seen to be a considerable reduction in the disparity between the eigenvalues of the system.

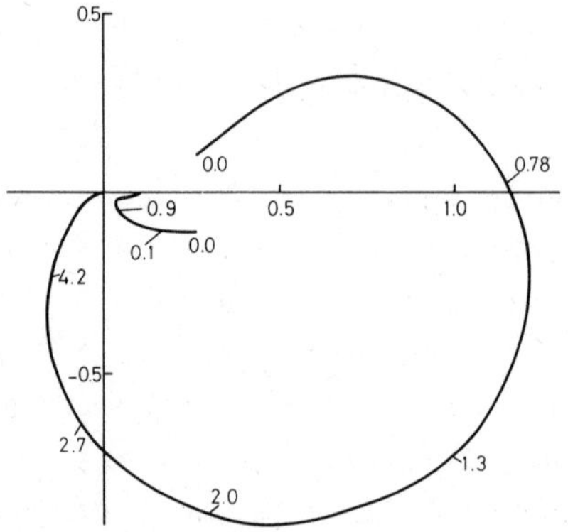

Fig. 4.5 Characteristic Loci of reactor with constant pre-compensator K_h to suppress high frequency interaction.

The larger locus in Figure 4.5 exhibits appreciable magnitude at frequencies of 3 and above, the range of discarded and neglected modes in the reactor model. Because of the uncertainties of the characteristic loci in this frequency range, one is not justified in designing on the basis of the loci at these frequencies. We therefore propose to attenuate and lag the large locus at frequencies in the region of $\omega = 2$ to bring its magnitude and phase into closer correspondence with the smaller locus.

Both the loci were lagged slightly in order to decrease the bandwidth to be within the range where the 7th order and 14th order models were similar. The compensators were introduced using the approximate commutative controller method calculated at a frequency $\omega = 2$,

$$K_1(s) = \begin{bmatrix} 0.852 & 0.620 \\ -0.414 & 0.923 \end{bmatrix} \begin{bmatrix} \frac{(0.1s+0.5)(0.5s+0.5)}{(s+0.5)(s+0.5} & 0 \\ 0 & \frac{2s+3}{s+0.5} \end{bmatrix}$$

$$\begin{bmatrix} 0.896 & -0.571 \\ 0.382 & 0.785 \end{bmatrix} \qquad (4.2)$$

The choice of the lag frequency 0.5 is influenced by the desire to decrease appreciably the magnitude of the locus at $\omega = 2$ but to maintain gains as high as possible at low and intermediate frequencies. The loci resulting from the introduction of $K_1(s)$ as a precompensator are shown in Figure 4.6 . The phases of the loci are now seen to be in much closer correspondence. The loci magnitudes are also more nearly equal.

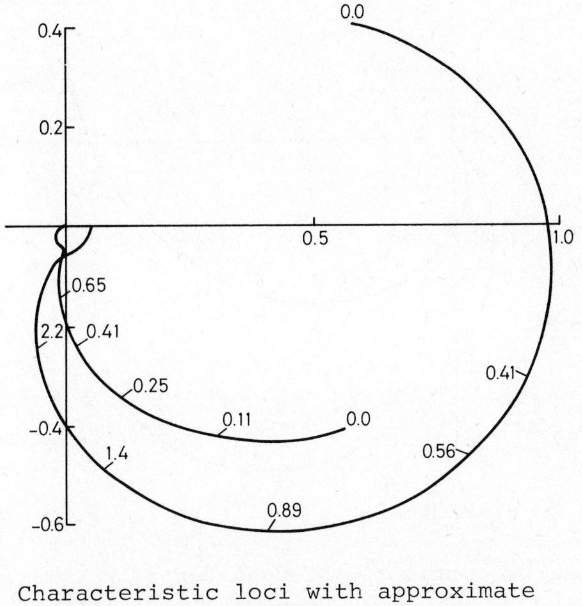

Figure 4.6 Characteristic loci with approximate commutative controller at = 2 . K_h and $K_1(s)$ given by equations (4.1) and (4.2)

Compensation is now introduced to further modify these loci for the purpose of boosting the magnitudes of each at intermediate and low frequencies without affecting the phase at high frequencies. Lags at $\omega = 0.2$ and 0.05, each with a

10-fold frequency range, are applied to the large and small loci respectively. These lags are introduced through an approximate commutative controller calculated at $\omega = 0.3$, the approximate centre of the frequency range of the two new lags. This precompensator is

$$K_2(s) = \begin{bmatrix} 1.223 & 0.810 \\ -0.446 & -0.651 \end{bmatrix} \begin{bmatrix} \frac{s+2.0}{s+0.2} & 0 \\ 0 & \frac{s+0.5}{s+0.05} \end{bmatrix} \begin{bmatrix} 1.495 & 1.858 \\ -1.023 & -2.807 \end{bmatrix}$$

The resulting loci are shown in Figure 4.7 . These loci seem balanced well enough to attempt closure of the loop.

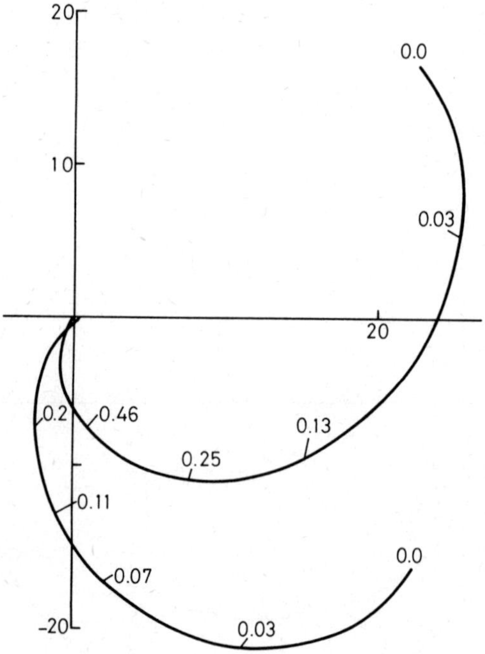

Fig. 4.7 Characteristic loci with K_h, $K_1(s)$ and $K_2(s)$.

A scalar proportional-integral controller is introduced at a frequency of 0.05; this produces high gains in both loci at low frequency with only a few degrees of phase shift in the slowest locus near the critical point. This compensator is taken to be

$$K_3(s) = 4\,\frac{s+0.1}{s}$$

The gain of 4 was selected to give a 45° phase margin to the locus having the largest phase lag near the critical point.

To test the performance of this cascade of compensators, transient responses to step changes in set point were calculated using both the 7th- and 14-th order reactor models. It was essential that the control system be tested on the higher order model because the reactor modes neglected in the design model could easily lead to instability if care had not been exercised in the design sequence. The closed-loop response of these two reactor models to a step increase in the setpoint C_{out} is shown in Figure 4.8 .

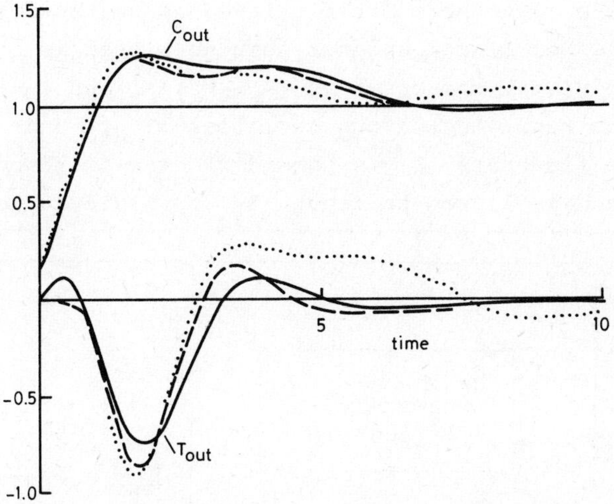

Fig. 4.8 Response of system to step changes in the setpoint for C_{out} .

—— 7th order model with controllers K_h, $K_1(s)$, $K_2(s)$, $K_3(s)$

--- 14th order model with controllers K_h $K_1(s)$, $K_2(s)$, $K_3(s)$

... 14th order model with controllers K_h, $K_1(s)$, $K_2(s)$, $K_3(s)$, $K_4(s)$.

The responses display an acceptable degree of stability;

however, there is considerable interaction present in the mid frequencies.

4.5 Design of the 2-Bed System

The incorporation of Bed I into the multivariable design just described for Bed II is accomplished in a simple way with simple objectives. Feed temperature T_f is the new input variable associated with Bed I and is known[8] to have little potential for regulating dynamic conditions in the reactor. Its sole function in this design is therefore relegated to driving the quench flow rate to its nominal value after the quench has accomplished the major portion of its short-term corrective action. The control system configuration can therefore be viewed as sketched in Figure 4.9 . The Bed II compensator just developed is the product of the individual compensators K_h, K_1, K_2 and K_3, and is employed to regulate the two variables T_{out} and C_{out}; the quench flow rate Q is considered the third output variable and is linked to input T_f by a single loop.

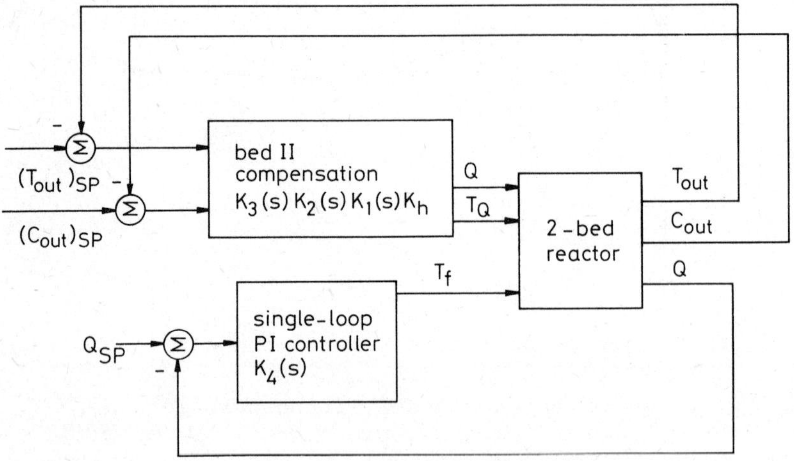

Fig. 4.9 Control System Configuration for 2-bed Reactor

The design of this simple loop between T_f and Q is straight forward. The Nyquist locus, which is shown in Figure 4.10, has a high-order character. The proportional-integral controller

$$K_4(s) = 0.07 \frac{s+0.2}{s}$$

when introduced into the loop yields a gain margin of about 3 and an integration rate that would reset Q in about 10 time units. Three time units correspond approximately to one "time constant" of the uncontrolled 2-bed reactor. In the course of the investigation of the performance of the complete system, it was found that some additional tuning could be made of the scalar PI controller $K_3(s)$ of bed II. The integration rate was doubled to give the modified scalar controller

$$K_3'(s) = 4 \frac{s+0.2}{s}$$

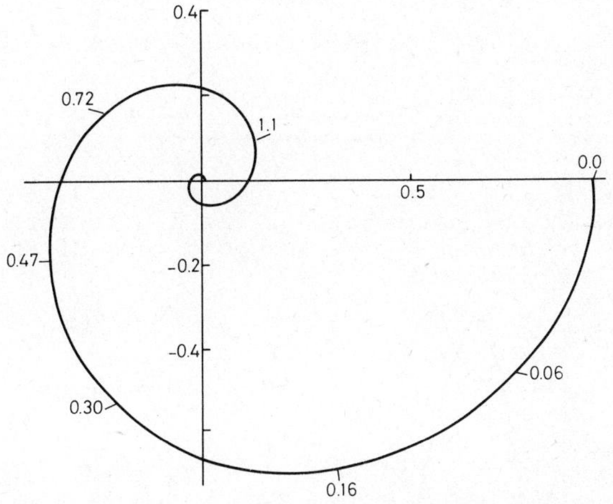

Fig. 4.10 Nyquist locus of $Q(s)/T_f(s)$ of Fig. 4.9, (bottom loop open, top loop closed).

4.6 System Performance and Properties

System performance was evaluated by simulation of the reactor under the controls devised above upon a step disturbance in the feed concentration C_o . Fourteenth-order models of both beds were used in these simulations.

A comparison of the controlled and uncontrolled transients in T_{out} and C_{out} is given in Figure 4.11 . Both temperature and concentration are seen to be held by the controls to small values relative to the uncontrolled case, and both are returned to the vicinity of the set point with reasonable speed. The control action required to achieve this performance is shown in Figure 4.12 . The response to a set point change is shown by the dotted line in Figure 4.8 . Mid-frequency interaction is evident in both set point and disturbance responses.

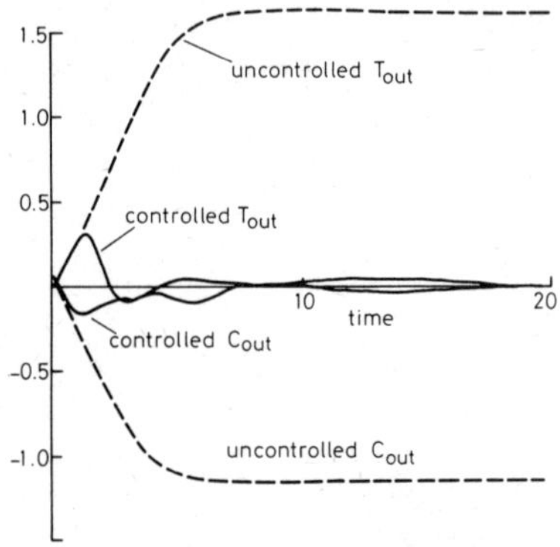

Fig. 4.11 Transient behaviour for a unit step disturbance on C_o using both beds of the 14th-order model.

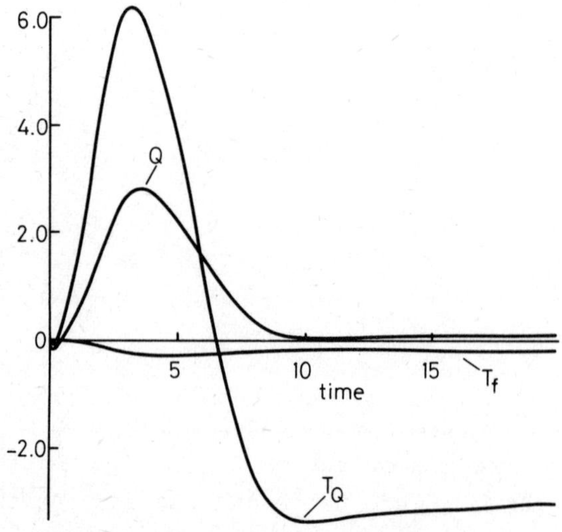

Fig. 4.12 Control action while suppressing a unit step disturbance on C_o for the 14th-order model.

The magnitude of the control inputs is an important matter in multivariable systems; it is certain to be larger than that of a single-output system because of the extra demands on the controller. For example, in this case independent control has been demanded over output temperature and concentration. In practice, the range of the inputs is limited and thus an equivalent concern is the magnitude of the disturbances that can be accommodated without saturation of the control inputs. The results of Figure 4.12 imply, for example, that this control system could accommodate feed concentration steps of up to 2.5% before saturation would occur on the quench flow rate of the experimental reactor used in earlier studies. For larger disturbances the transient interaction would increase, but steady-state regulation could still be achieved until one of the heaters saturates. In the single-output systems of Silva and Wallman concentration disturbances of 10 to 20% could be handled before the quench flow rate reached its limits.

A further design study using a computer to tune the control scheme[3] reduced the use of Q by a factor of 3 while also reducing the interaction considerably. This would allow disturbances of up to 7.5% before Q saturates, which is much more comparable with the single output results.

The three characteristic loci of the final 3x3 system are shown in Figure 4.13 . One of the three is considerably slower than the other two and is undoubtedly associated with the more or less independent PI controller linking Q to T_f . The two faster loci exhibit gain and phase characteristics near the critical point that would contribute to some slight overshoot of the controlled variables, and indeed this is observed in the response of T_{out} and C_{out} shown in Figure 4.11 . Quench flow rate Q, however, responds in an overdamped manner, as would be suggested by the shape of the slower locus.

It is noticed that the seven poles introduced through the compensators $K_1(s)$, $K_2(s)$, $K_3(s)$ and $K_4(s)$, are clustered in the region 0 to -0.5, about four times slower than the reactor poles. Further, there is an excess of poles over zeros introduced in that region. This compensator pole-zero configuration is typical of that usually used for high-order cascaded processes.

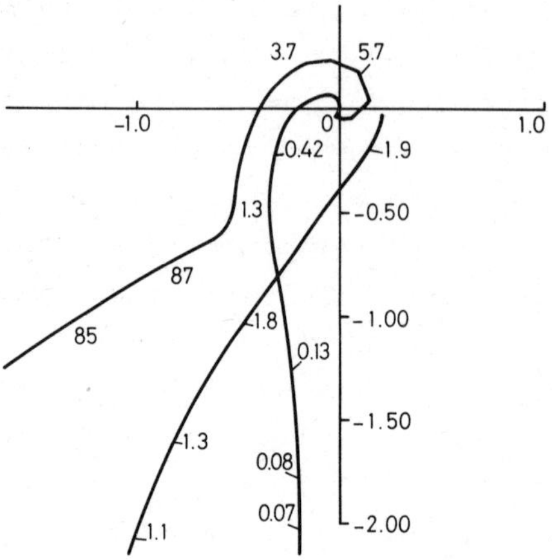

Fig. 4.13 Characteristic Loci for Final Design.

The implementation of the set of compensators would be straightforward in either analogue or digital realizations. In a digital implementation, a 7th-order state-space realization of the compensator transfer functions would be required, a size that is easily accommodated in a process control computer. More details of this design study can be found in reference 4.

Appendix

Matrices for the Reactor System Model

Time unit	=	87.5 sec
Concentration unit	=	1.0 mole %
Temperature unit	=	167.4 ^{o}C
Flow rate unit	=	13.5 ℓ/min.

Feed heater $\frac{1}{\tau} = 0.2956$

Bed 1

$$A^{I} = \begin{bmatrix} -2.3832 & -0.1717 & 0.1562 & -0.1606 & 0.1802 & -0.2527 & 0.1550 \\ 0.9280 & -1.8688 & -0.3027 & 0.2642 & -0.2818 & 0.3885 & -0.2375 \\ 0.5462 & 1.8614 & -1.4986 & -0.4074 & 0.3689 & -0.4846 & 0.2938 \\ -0.1813 & 0.4976 & 2.3399 & -1.4191 & -0.4404 & 0.4899 & -0.2892 \\ -0.3797 & -0.7642 & 0.6631 & 2.5293 & -1.7368 & -0.3369 & 0.1716 \\ -0.0805 & -0.8628 & -0.6172 & 1.7163 & 2.0183 & -2.2067 & -0.1147 \\ 0.0265 & -0.7599 & -0.8771 & 1.4123 & 2.0361 & 0.8079 & -2.7725 \end{bmatrix}$$

$$B^{I} = \begin{bmatrix} 2.5936 & 0.0179 \\ 1.5082 & 0.0645 \\ -0.1333 & 0.0946 \\ -0.5473 & 0.0902 \\ 0.2840 & 0.1007 \\ 0.6519 & 0.1258 \\ 0.6540 & 0.1356 \end{bmatrix}, \quad D^{I} = \begin{bmatrix} 0.1760 & 0.0000 \\ -0.1088 & 0.0000 \\ 0.0450 & 0.0000 \\ 0.0027 & 0.0000 \\ 0.0000 & 0.7789 \end{bmatrix}$$

$$C^{I} = \begin{bmatrix} -0.3755 & 0.8200 & 0.4834 & -0.1698 & 0.1229 & -0.1427 & 0.0840 \\ 0.2118 & -0.2439 & 0.5766 & 0.7193 & -0.2745 & 0.2758 & -0.1585 \\ -0.0852 & 0.0860 & -0.1199 & 0.3097 & 0.9534 & -0.4021 & 0.2112 \\ -0.0050 & 0.0048 & -0.0059 & 0.0098 & -0.0259 & 0.2917 & 0.7260 \\ -0.0846 & -0.1912 & -0.2804 & -0.3298 & -0.3012 & -0.1621 & 0.0000 \end{bmatrix}$$

Mixing Chamber

$$M = \begin{bmatrix} -0.3373 & 0.2370 & 0.7630 & 0.0000 \\ -0.7226 & 0.0000 & 0.0000 & 0.7630 \\ 1.0000 & 0.0000 & 0.0000 & 0.0000 \end{bmatrix}$$

Bed 11

$$A^{II} = \begin{bmatrix} -2.5108 & -0.1285 & 0.1029 & -0.1277 & 0.1581 & -0.2207 & 0.1346 \\ 0.7675 & -2.1256 & -0.2003 & 0.2004 & -0.2335 & 0.3194 & -0.1940 \\ 0.8449 & 1.7995 & -1.8809 & -0.2849 & 0.2694 & -0.3464 & 0.2080 \\ 0.3177 & 1.5039 & 2.4101 & -1.8352 & -0.2749 & 0.2818 & -0.1626 \\ -0.6847 & -0.2184 & 1.6907 & 2.2016 & -2.0525 & -0.2373 & 0.1205 \\ -1.0197 & -1.4581 & 0.3344 & 1.9009 & 1.4282 & -2.3437 & -0.1116 \\ -1.0121 & -1.6634 & -0.0508 & 1.7249 & 1.4403 & 0.6234 & -2.8285 \end{bmatrix}$$

$$B^{II} = \begin{bmatrix} 2.8314 & 0.0679 & -0.0266 \\ 2.3639 & 0.2812 & -0.1235 \\ 0.8489 & 0.5514 & -0.2026 \\ -1.4594 & 0.5891 & -0.0781 \\ -1.5940 & 0.3488 & 0.2194 \\ -0.1837 & 0.2054 & 0.3809 \\ 0.2538 & 0.1872 & 0.4044 \end{bmatrix}, \quad D^{II} = \begin{bmatrix} 0.0471 & 0.0000 & -0.0029 \\ -0.0941 & 0.0000 & -0.0039 \\ 0.0609 & 0.0000 & -0.0035 \\ 0.0026 & 0.0000 & -0.0019 \\ 0.0000 & 0.2433 & 0.2759 \end{bmatrix}$$

$$C^{II} = \begin{bmatrix} -0.1104 & 1.0098 & 0.0706 & -0.0319 & 0.0244 & -0.0289 & 0.0171 \\ 0.1903 & -0.2400 & 0.8659 & 0.3817 & -0.1912 & 0.2017 & -0.1168 \\ -0.1188 & 0.1262 & -0.1865 & 0.6320 & 0.6820 & -0.4327 & 0.2346 \\ -0.0049 & 0.0049 & -0.0060 & 0.0100 & -0.0263 & 0.3416 & 0.6759 \\ -0.0465 & -0.1135 & -0.1909 & -0.2619 & -0.2634 & -0.1422 & -0.0002 \end{bmatrix}$$

References

1. EDMUNDS, J.M. : "The Cambridge Linear Analysis and Design Program", CUED/F-CAMS/TR 170, Eng. Dept., University of Cambridge, (1978).

2. EDMUNDS, J.M. and KOUVARITAKIS, B. : "Multivariable root loci : a unified approach to finite and infinite zeros", Int. J. Control, vol. 29 No. 3 (1979).

3. EDMUNDS, J.M. : "Control system design and analysis using closed-loop Nyquist and Bode arrays", Int. J. Control, vol. 30, No. 5, (1979), 773.

4. FOSS, A.S., EDMUNDS, J.M. and KOUVARITAKIS, B. : "Multivariable control system for two-bed reactors by the characteristic locus method", Ind. Eng. Chem. Fund. 19, (1980), 109.

5. MACFARLANE, A.G.J. and KOUVARITAKIS, B. : "A design technique for linear multivariable feedback systems", Int. J. Control, vol. 25, No. 6, (1977), 837.

6. MICHELSEN, M.L., VAKIL, H.B., FOSS, A.S., Ind. Engr. Chem. Fund., 12 (1973), 323.

7. SILVA, J.M. : Ph.D. Thesis, University of California, Berkeley, (1978).

8. SILVA, J.M., WALLMAN, P.H., FOSS, A.S. : Ind. Eng. Chem. Fund, 18, (1979), 383.

9. WALLMAN, P.H. : Ph.D. Thesis, University of Calif., Berkeley, (1977).

10. WALLMAN, P.H., SILVA, J.M., FOSS, A.S. : Ind. Eng. Chem Fund, 18, (1979), 392.

Chapter 5

The inverse Nyquist array design method

Professor N. MUNRO

Synopsis

The general problem of the design of multivariable control systems is considered and the stability of multivariable feedback systems is examined. The concept of 'diagonal dominance' is introduced, and Rosenbrock's Inverse Nyquist Array Design Method is developed. Methods of achieving diagonal dominance are discussed and illustrated in terms of practical problems.

5.1 Introduction

The design of control systems for single-input single-output plant using the classical frequency response methods of Bode, Nyquist and Nichols is well established. However, the frequency response approach of Nyquist has been extended by Rosenbrock[1] to deal with multi-input multi-output plant where significant interaction is present.

During the last decade, interactive computing facilities have developed rapidly and it is now possible to communicate with a digital computer in a variety of ways; e.g. graphic display systems with cursors, joysticks, and light-pens. Equally, the digital computer can present information to the user in the form of graphs on a display terminal or as hard-copy on a digital plotter. The classical frequency-response methods for single-input single-output systems rely heavily on graphical representations, and Rosenbrock's 'inverse' Nyquist array' design method for multivariable systems suitably exploits the graphical output capabilities of the modern digital computer system. Also, the increased complexity of multivariable systems has made it necessary to employ interactive computer-aided design facilities, such as those developed at the Control Systems Centre, UMIST, Manchester[2], in order to establish an effective dialogue with the user.

5.2 The Multivariable Design Problem

In general, a multivariable control system will have m inputs and ℓ outputs and the system can be described by an $\ell \times m$ transfer function matrix G(s). Since we are interested in feedback control we almost always have $\ell = m$.

The uncontrolled plant is assumed to be described by an rational transfer function matrix G(s) and we wish to determine a controller matrix K(s) such that when we close the feedback loops through the feedback matrix F, as in Figure 5.1, the system is stable and has suitably fast responses.

The matrix F is assumed diagonal and independent of s, i.e.

$$F = \text{diag}\{f_i\} \tag{5.1}$$

F represents loop gains which will usually be implemented in the forward path, but which it is convenient to move into the return path. In addition, the design will have high integrity if the system remains stable as the gain in each loop is reduced.

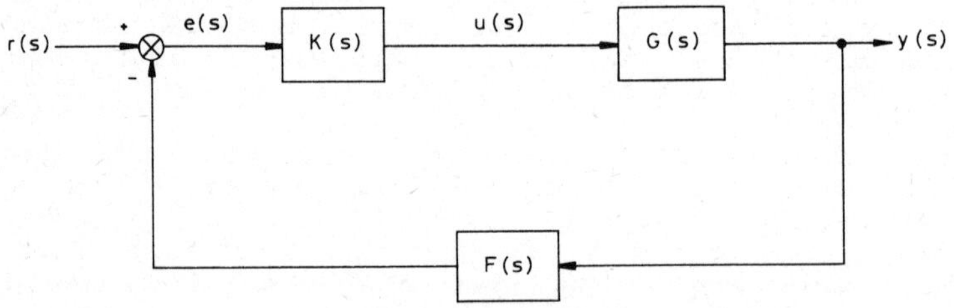

Fig. 5.1

Let us consider the controller matrix K(s) to consist of the product of two matrices $K_p(s)$ and K_d, i.e.

$$K(s) = K_p(s)\, K_d \tag{5.2}$$

where K_d is a diagonal matrix independent of s; i.e.

$$K_d = \text{diag}\{k_i\}, \quad i = 1,\ldots,m \tag{5.3}$$

Then the system of Figure 5.1 can be re-arranged as shown in Figure 5.2 where the stability of the closed-loop system is unaffected by the matrix K_d outside the loop. For convenience, we rename the matrix K_dF as F. All combinations of gains and of open or closed-loops can now be obtained by a suitable choice of the f_i, and we shall want the system to remain stable for all values of the f_i from zero up to their design values.

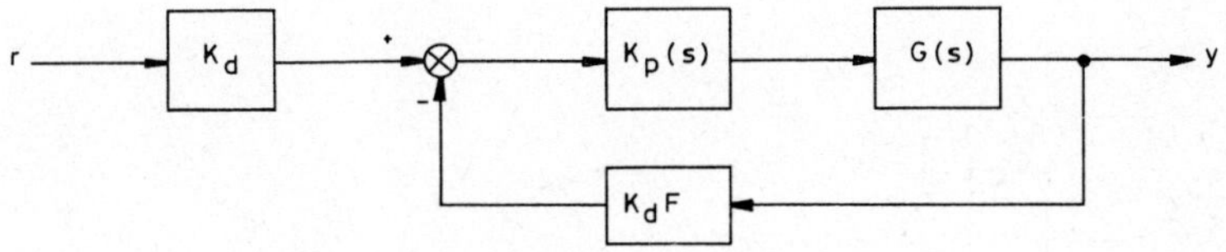

Fig. 5.2

When $f_1 = 0$ the first loop is open, and all gains in the first loop up to the design value f_{1d} can be achieved by increasing f_1.

The elements f_i of $F = \text{diag}\{f_i\}$ can be represented by points in an m-dimensional space which can be called the *gain space*. That part of the gain space in which $f_i > 0$, $i = 1,\ldots,m$ corresponds to negative feedback in all loops, and is the region of most practical interest. The point $\{f_1, f_2, \ldots, f_m\}$ belongs to the asymptotically stable region in the gain space if and only if the system is asymptotically stable with $F = \text{diag}\{f_i\}$.

Let

$$Q(s) = G(s)K(s) \tag{5.4}$$

then the closed-loop system transfer function matrix $H(s)$ is given by

$$H(s) = (I+Q(s)F)^{-1}Q(s) = Q(s)(I+FQ(s))^{-1} \quad (5.5)$$

Consider the open-loop system

$$Q(s) = \begin{pmatrix} \frac{1}{s+1} & \frac{2}{s+3} \\ \frac{1}{s+1} & \frac{1}{s+1} \end{pmatrix} \quad (5.6)$$

Then if $f_1 = 10$ and $f_2 = 0$, $H(s)$ has all of its poles in the open left-half plane and the system is asymptotically stable. However, for $f_1 = 10$ and $f_2 = 10$, the closed-loop system $H(s)$ is unstable, as shown in Figure 5.3 .

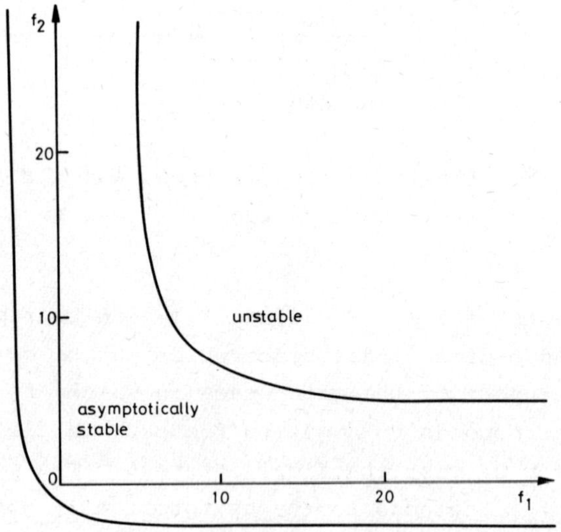

Fig. 5.3

This situation could have been predicted by examining the McMillan form of $Q(s)$ which is

$$M(s) = \begin{pmatrix} \frac{1}{(s+1)(s+3)} & 0 \\ 0 & \frac{(s-1)}{(s+1)} \end{pmatrix} \tag{5.7}$$

where the poles of the McMillan form are referred to as the "poles of the system" and the zeros of the McMillan form are referred to as the "zeros of the system". If any of the "zeros" of $Q(s)$ lie in the closed right-half plane, then it will not be possible to set up m high gain control loops around this system.

Consider now the gain space for the system described by

$$Q(s) = \begin{pmatrix} \frac{s-1}{(s+1)^2} & \frac{5s+1}{(s+1)^2} \\ \frac{-1}{(s+1)^2} & \frac{s-1}{(s+1)^2} \end{pmatrix} \tag{5.8}$$

which is shown in Figure 5.4 . Now, up to a certain point, increase of gain in one loop allows increase of gain in the other loop without instability. The McMillan form of this latter system $Q(s)$ is

$$M(s) = \begin{pmatrix} \frac{1}{(s+1)^2} & 0 \\ 0 & \frac{s+2}{s+1} \end{pmatrix} \tag{5.9}$$

which implies that, despite the non-minimum phase terms in the diagonal elements of $Q(s)$, no non-minimum phase behaviour will be obtained with the feedback loops closed. It is also interesting to note that this system is stable with a small amount of positive feedback. However, this is not a high integrity system, since failure of any one loop may put the system into an unstable region of operation.

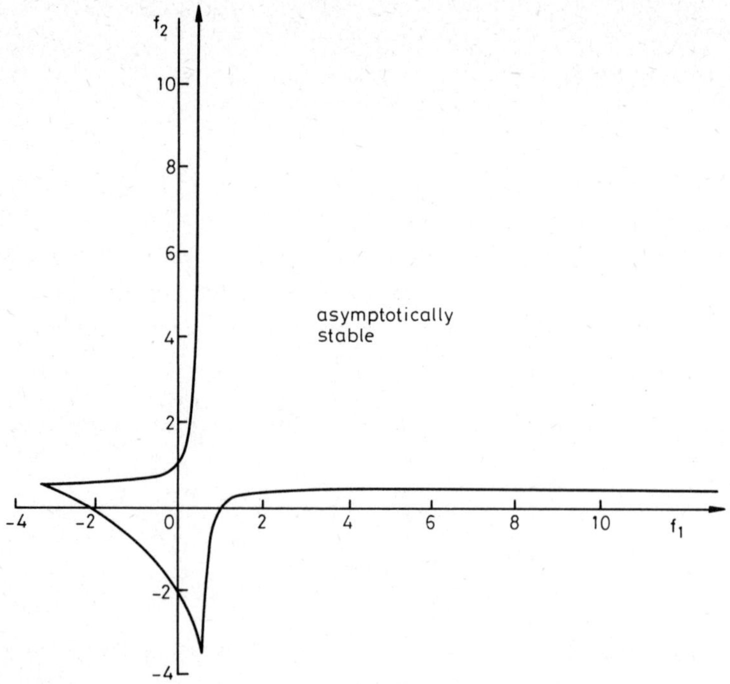

Fig. 5.4

5.3 Stability

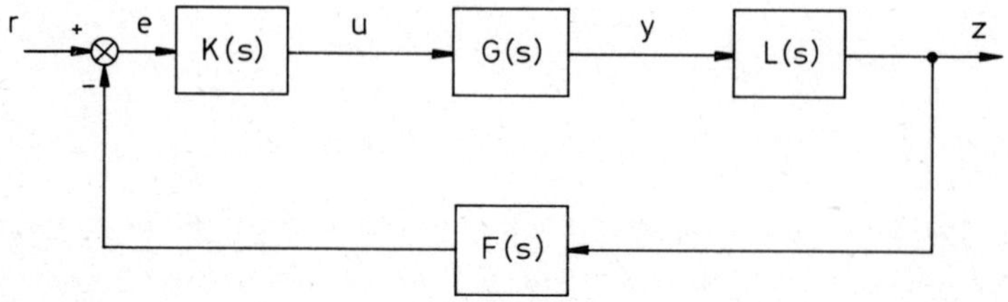

Fig. 5.5

Consider the system shown in Figure 5.5, in which all matrices are mxm; this last condition is easily relaxed (see Reference 1, p. 131 ff). Let $Q = LGK$, and suppose that G arises from the system matrix

$$P_G(s) = \begin{pmatrix} T_G(s) & U_G(s) \\ -V_G(s) & W_G(s) \end{pmatrix} \tag{5.10}$$

and similarly for K(s), L(s), F(s). For generality, we do not require any of these system matrices to have least order. The equations of the closed-loop system can then be written as

$$\left(\begin{array}{cccccccc|c} T_K & U_K & 0 & 0 & 0 & 0 & 0 & 0 & 0 \\ -V_K & W_K & 0 & -I_m & 0 & 0 & 0 & 0 & 0 \\ 0 & 0 & T_G & U_G & 0 & 0 & 0 & 0 & 0 \\ 0 & 0 & -V_G & W_G & 0 & -I_m & 0 & 0 & 0 \\ 0 & 0 & 0 & 0 & T_L & U_L & 0 & 0 & 0 \\ 0 & 0 & 0 & 0 & -V_L & W_L & 0 & -I_m & 0 \\ 0 & 0 & 0 & 0 & 0 & 0 & T_F & U_F & 0 \\ 0 & I_m & 0 & 0 & 0 & 0 & -V_F & W_F & -I_m \\ \hline 0 & 0 & 0 & 0 & 0 & 0 & 0 & I & 0 \end{array}\right)\left(\begin{array}{c} \bar{\xi}_K \\ -\bar{e} \\ \bar{\xi}_G \\ -\bar{u} \\ \bar{\xi}_L \\ -\bar{y} \\ \bar{\xi}_F \\ -\bar{z} \\ \hline -\bar{v} \end{array}\right) = \left(\begin{array}{c} 0 \\ 0 \\ 0 \\ 0 \\ 0 \\ 0 \\ 0 \\ 0 \\ \hline -\bar{z} \end{array}\right) \qquad (5.11)$$

Here the matrix on the left-hand side of equation (5.11) is a system matrix for the closed-loop system shown in Figure 5.5, which we can also write as

$$P_H = \begin{pmatrix} T_H & U_H \\ -V_H & W_H \end{pmatrix} \qquad (5.12)$$

The closed-loop system poles are the zeros of $|T_H(s)|$, and Rosenbrock has shown[1] that

$$|T_H(s)| = |I_m+Q(s)F(s)|\,|T_L(s)|\,|T_G(s)|\,|T_K(s)|\,|T_F(s)| \qquad (5.13)$$

As the zeros of $|T_L||T_G||T_K||T_F|$ are the open-loop poles, we need only the following information to investigate stability,

(i) The rational function $|I_m+Q(s)F(s)|$
(ii) The locations of any open-loop poles in the

closed right half-plane (crhp).

Notice that this result holds whether or not the subsystems have least order. If we apply Cauchy's theorem, (ii) can be further reduced: all we need is the number p_o of open-loop poles in the crhp.

From equation (5.5)

$$|I_m+QF| = |Q(s)|/|H(s)| \tag{5.14}$$

The Nyquist criterion depends on encirclements of a critical point by the frequency response locus of the system to indicate stability. Let D be the usual Nyquist stability contour in the s-plane consisting of the imaginary axis from -jR to +jR, together with a semi-circle of radius R in the right half plane. The contour D is supposed large enough to enclose all finite poles and zeros of $|Q(s)|$ and $|H(s)|$, lying in the closed right half plane.

Let $|Q(s)|$ map D into Γ_Q, while $|H(s)|$ maps D into Γ_H. As s goes once clockwise around D, let Γ_Q encircle the origin N_Q times clockwise, and let Γ_H encircle the origin N_H times clockwise. Then, if the open-loop system characteristic polynomial has p_o zeros in the closed right half plane, the closed-loop system is asymptotically stable if and only if

$$N_H - N_Q = p_o \tag{5.15}$$

This form of the stability theorem is directly analogous to the form used with single-input single-output systems, but is difficult to use since $|H(s)|$ is a complicated function of $Q(s)$, namely

$$|H(s)| = |Q(s)|/|I+QF| \tag{5.16}$$

Equation (5.5),

$$H(s) = (I+Q(s)F)^{-1}Q(s)$$

shows that the relationship between the open-loop system

$Q(s)$ and the closed-loop system $H(s)$ is not simple. However, if $Q^{-1}(s)$ exists then

$$H^{-1}(s) = F + Q^{-1}(s) \tag{5.17}$$

which is simpler to deal with. Instead of $H^{-1}(s)$ and $Q^{-1}(s)$ we shall write $\hat{H}(s) = H^{-1}(s)$ and $\hat{Q}(s) = Q^{-1}(s)$. Then, the $\hat{h}_{ii}(s)$ are the diagonal elements of $H^{-1}(s)$. In general, $\hat{h}_{ii}(s) \neq h_{ii}^{-1}(s)$, where $h_{ii}^{-1}(s)$ is the inverse of the diagonal element $h_{ii}(s)$ of $H(s)$.

We shall, in what follows, develop the required stability theorems in terms of the inverse system matrices. Also, we note that if $K(s)$ has been chosen such that $Q(s) = G(s)K(s)$ is diagonal and if F is diagonal, then we have m single loops. However, several objections to this approach can be made. In particular, it is an unnecessary extreme. Instead of diagonalising the system, we shall consider the much looser criterion of diagonal dominance.

5.3.1 Diagonal dominance[1]

A rational mxm matrix $\hat{Q}(s)$ is *row diagonal dominant* on D if

$$|\hat{q}_{ii}(s)| > \sum_{\substack{j=1 \\ j\neq i}}^{m} |\hat{q}_{ij}(s)| \tag{5.18a}$$

for i = 1,...,m and all s on D. *Column diagonal dominance* is defined similarly by

$$|\hat{q}_{ii}(s)| > \sum_{\substack{j=1 \\ j\neq i}}^{m} |\hat{q}_{ji}(s)| \tag{5.18b}$$

The dominance of a rational matrix $\hat{Q}(s)$ can be determined by a simple graphical construction. Let $\hat{q}_{ii}(s)$ map D into $\hat{\Gamma}_i$ as in Figure 5.6 . This will look like an inverse Nyquist plot, but does not represent anything directly measurable on the physical system. For each s on D draw a circle of radius

$$d_i(s) = \sum_{\substack{j=1 \\ j \neq i}}^{m} |\hat{q}_{ij}(s)| \qquad (5.19)$$

centred on the appropriate point of $\hat{q}_{ii}(s)$, as in Figure 5.6 . Do the same for the other diagonal elements of $\hat{Q}(s)$. If each of the bands so produced excludes the origin, for $i = 1,\ldots,m$, then $\hat{Q}(s)$ is row dominant on D. A similar test for column dominance can be defined by using circles of radius

$$d_i'(s) = \sum_{\substack{j=1 \\ j \neq i}}^{m} |\hat{q}_{ji}(s)| \qquad (5.20)$$

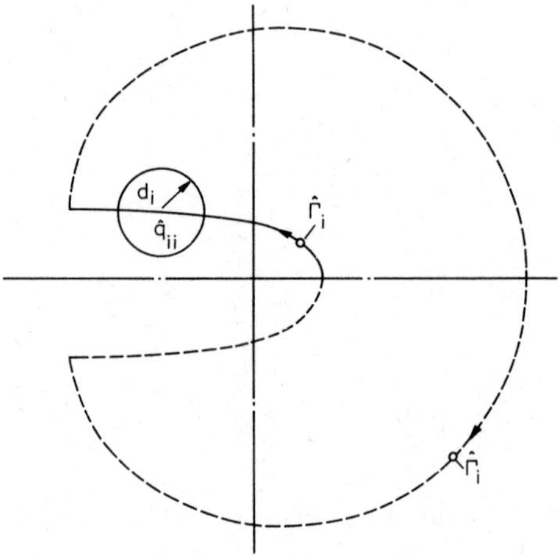

Fig. 5.6

5.3.2 Further stability theorems

If $\hat{Q}(s)$ is row (or column) dominant on D, having on it no zero of $|\hat{Q}(s)|$ and no pole of $\hat{q}_{ii}(s)$, for $i = 1,\ldots,m$, then let $\hat{q}_{ii}(s)$ map D into $\hat{\Gamma}_i$ and $|\hat{Q}(s)|$ map D into $\hat{\Gamma}_Q$. If $\hat{\Gamma}_i$ encircles the origin $\hat{N}_i$ times and $\hat{\Gamma}_Q$ encircles the origin $\hat{N}_Q$ times, all encirclements being clockwise, then Rosenbrock[1] has shown that

$$\hat{N}_Q = \sum_{i=1}^{m} \hat{N}_i \tag{5.21}$$

A proof of this result is given in the Appendix.

Let $\hat{Q}(s)$ and $\hat{H}(s)$ be dominant on D, let $\hat{q}_{ii}(s)$ map D into $\hat{\Gamma}_{qi}$ and let $\hat{h}_{ii}(s)$ map D into $\hat{\Gamma}_{hi}$. Let these encircle the origin $\hat{N}_{qi}$ and $\hat{N}_{hi}$ times respectively. Then with p_o defined as in (5.15), the closed-loop system is asymptotically stable if, and only if,

$$\sum_{i=1}^{m} \hat{N}_{qi} - \sum_{i=1}^{m} \hat{N}_{hi} = p_o \tag{5.22}$$

This expression represents a generalised form of Nyquist's stability criterion, applicable to multivariable systems which are diagonal dominant.

$$\text{For } |Q|, \quad \hat{N}_Q = - \sum_{i=1}^{m} \hat{N}_{qi} \tag{5.23}$$

and since we are considering inverse polar plots stability is determined using

$$\hat{N}_H - \hat{N}_Q = - p_o + p_c \tag{5.24}$$

where $p_c = 0$ for closed-loop stability.

Hence, $$\hat{N}_Q - \hat{N}_H = p_o \tag{5.25}$$

replaces the relationship used with direct polar plots, given by equation (5.15).

5.3.3 Graphical criteria for stability

If for each diagonal element $\hat{q}_{ii}(s)$, the band swept out by its circles does not include the origin or the critical point $(-f_i, 0)$, and if this is true for $i = 1, 2, \ldots, m$, then the generalised form of the inverse Nyquist stability criterion, defined by (5.25), is satisfied. In general, the $\hat{q}_{ii}(s)$ do not represent anything directly measurable on the system. However, using a theorem due to Ostrowski[3]

$$h_i^{-1}(s) = h_{ii}^{-1}(s) - f_i \tag{5.26}$$

is contained within the band swept out by the circles centred on $\hat{q}_{ii}(s)$, and this remains true for all values of gain f_i in each other loop j between zero and f_{jd}. Note that $h_{ii}^{-1}(s)$ is the inverse transfer function seen between input i and output i with all loops closed. The transfer function $h_i(s)$ is that seen in the ith loop when this is open, but the other loops are closed. It is this transfer function for which we must design a single-loop controller for the ith loop.

The theorems above tell us that as the gain in each other loop changes as long as dominance is maintained in the other loops, $h_{ii}^{-1}(s)$ will also change but always remains inside the ith Gershgorin band. The band within which $h_i^{-1}(s)$ lies can be further narrowed. If $\hat{Q}$ and $\hat{H}$ are dominant, and if

$$\phi_i(s) = \max_{\substack{j \\ j\neq i}} \frac{d_j(s)}{|f_j+\hat{q}_{jj}(s)|} \tag{5.27}$$

then $h_i^{-1}(s)$ lies within a band based on $\hat{q}_{ii}(s)$ and defined by circles of radius

$$r_i(s) = \phi_i(s)d_i(s) \tag{5.28}$$

Thus, once the closed-loop system gains have been chosen such that stability is achieved in terms of the larger bands, then a measure of the gain margin for each loop can be determined by drawing the smaller bands, using the 'shrinking factors' $\phi_i(s)$ defined by (5.27), with circles of radius r_i. These narrower bands, known as Ostrowski bands, also reduce the region of uncertainty as to the actual location of the inverse transfer function $h_{ii}^{-1}(s)$ for each loop.

5.4 Design Technique

Using the ideas developed in the previous section, the frequency domain design method proposed by Rosenbrock consists essentially of determing a matrix $K_p(s)$ such that the product $[G(s)K_p(s)]^{-1}$ is diagonal dominant. When this condition has been achieved then the diagonal matrix $K_d(s)$ can be used to implement single-loop compensators as required to meet the overall design specification. Since the design is carried out using the inverse transfer function matrix then we are essentially trying to determine an inverse pre-compensator $\hat{K}_p(s)$ such that $\hat{Q}(s) = \hat{K}_p(s)\hat{G}(s)$ is diagonal dominant. The method is well suited to interactive graphical use of a computer.

One method of determining $\hat{K}_p(s)$ is to build up the required matrix out of elementary row operations using a graphical display of all of the elements of $\hat{Q}(s)$ as a guide. This approach has proven successful in practice and has, in most cases considered to-date, resulted in $K_p(s)$ being a simple matrix of real constants which can be readily realized.

Another approach which has proved useful is to choose $\hat{K}_p = G(o)$, if $|G(o)|$ is nonsingular. Here again $\hat{K}_p(s)$ is a matrix of real constants which simply diagonalizes the plant at zero frequency.

For example, Figure 5.7 shows the inverse Nyquist array (INA) of an uncompensated system with 2 inputs and 2 outputs.

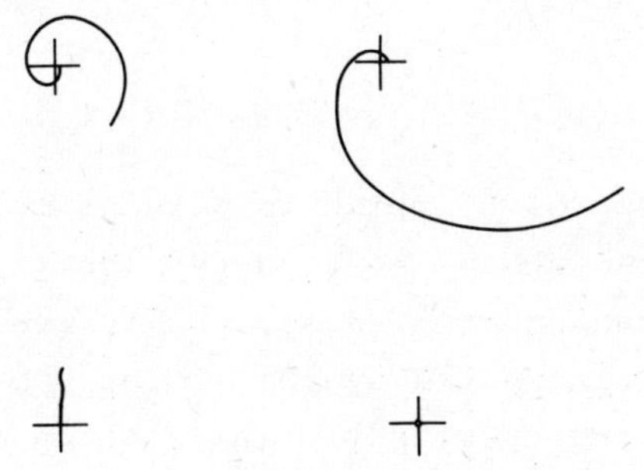

Fig. 5.7

An inspection of this diagram shows that element 1,2 is larger than element 1,1 at the maximum frequency considered, and similarly element 2,1 is very much larger than element 2,2 over a wide range of frequencies . Thus, the open-loop system is not row diagonal dominant, nor is it column diagonal dominant. However, by choosing $\hat{K}_p = G(o)$ the resulting INA of $\hat{Q}(j\omega)$ is shown in Figure 5.8 with the Gershgorin circles for row dominance superimposed. Both $\hat{q}_{11}$ and $\hat{q}_{22}$ are made diagonal dominant with this simple operation. The dominance of element $\hat{q}_{22}$ of this array can be further improved by another simple row operation of the form

$$R_2' = R_2 + \alpha R_1 \qquad (5.29)$$

where $\alpha \simeq 0.5$, in this case.

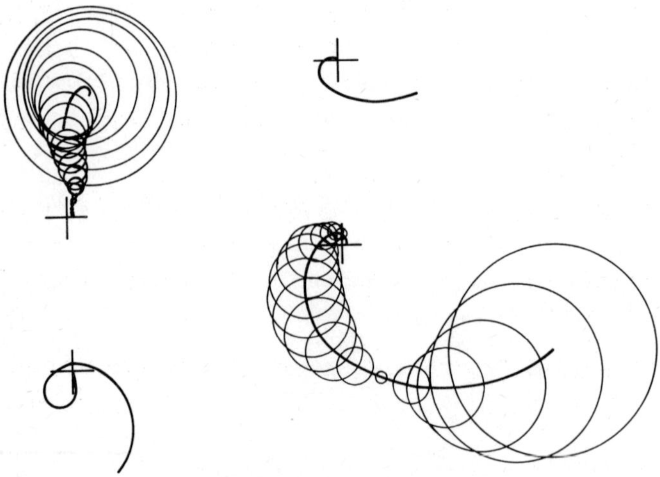

Fig. 5.8

A further approach, which is perhaps more systematic than those mentioned above, is to determine $\hat{K}_p$ as the 'best', in a least-mean-squares sense, wholly real matrix which most nearly diagonalizes the system $\hat{Q}$ at some frequency $s = j\omega$, (Rosenbrock[1] and Hawkins[4]). This choice of $\hat{K}_p$ can be considered as the best matrix of real constants which makes the sum of the moduli of the off-diagonal elements in each row of $\hat{Q}$ as small as possible compared with the modulus of the diagonal element at some frequency $s = j\omega$. The development of the two forms of this latter approach is given in the following paragraphs.

Consider the elements $\hat{q}_{jk}$ in some row j of $\hat{Q}(j\omega) = \hat{K}\hat{G}(j\omega)$, i.e.

$$\hat{q}_{jk}(j\omega) = \sum_{i=1}^{m} \hat{k}_{ji}\, \hat{g}_{ik}(j\omega) \tag{5.30}$$

$$= \sum_{i=1}^{m} \hat{k}_{ji}(\alpha_{ik}+j\beta_{ik}) \tag{5.31}$$

Now choose $\hat{k}_{j1}, \hat{k}_{j2}, \ldots, \hat{k}_{jm}$ so that

$$\sum_{\substack{k=1\\k\neq j}}^{m} |\hat{q}_{jk}(j\omega)|^2 \tag{5.32}$$

is made as small as possible subject to the constraint that

$$\sum_{i=1}^{m} \hat{k}_{ji}^2 = 1 \tag{5.33}$$

Using a Lagrange multiplier, we minimize

$$\phi_j = \sum_{\substack{k=1\\k\neq j}}^{m} \left| \sum_{i=1}^{m} \hat{k}_{ji}(\alpha_{ik}+j\beta_{ik}) \right|^2 + \lambda\left(\quad - \sum_{i=1}^{m} \hat{k}_{ji}^2 \right) \tag{5.34}$$

$$= \sum_{\substack{k=1\\k\neq j}}^{m} \left(\left(\sum_{i=1}^{m} \hat{k}_{ji}\alpha_{ik} \right)^2 + \left(\sum_{i=1}^{m} \hat{k}_{ji}\beta_{ik} \right)^2 + \lambda\left(1 - \sum_{i=1}^{m} \hat{k}_{ji}^2\right) \right) \tag{5.35}$$

and taking partial derivatives of ϕ_j with respect to $\hat{k}_{j\ell}$ (i.e. the elements of the row vector $\hat{k}_j$), we get, on setting these equal to zero,

$$\frac{\partial \phi_j}{\partial \hat{k}_{j\ell}} = \sum_{\substack{k=1\\k\neq j}}^{m} \left(2\left(\sum_{i=1}^{m} \hat{k}_{ji}\alpha_{ik} \right)\alpha_{\ell k} + 2\left(\sum_{i=1}^{m} \hat{k}_{ji}\beta_{ik} \right)\beta_{\ell k} \right) - 2\lambda\hat{k}_{j\ell} \tag{5.36}$$

$$= 0 \quad \text{for } \ell = 1,2,\ldots,m$$

Now writing

$$A_j = \{a_{i\ell}^{(j)}\}$$

$$= \left[\sum_{\substack{k=1 \\ k \neq j}}^{m} (\alpha_{ik}\alpha_{\ell k} + \beta_{ik}\beta_{\ell k}) \right] \tag{5.37}$$

and

$$\hat{k}_j = (\hat{k}_{j\ell}) \tag{5.38}$$

the minimization becomes

$$A_j\hat{k}_j^T - \lambda\hat{k}_j^T = 0 \tag{5.39}$$

since

$$\sum_{\substack{k=1 \\ k \neq j}}^{m} |\hat{q}_{jk}(j\omega)|^2 = \hat{k}_j A_j \hat{k}_j^T$$

$$= \lambda\hat{k}_j\hat{k}_j^T$$

$$= \lambda \tag{5.40}$$

Thus, the design problem becomes an eigenvalue/eigenvector problem where the row vector $\hat{k}_j$, which pseudodiagonalizes row j of $\hat{Q}$ at some frequency $j\omega$, is the eigenvector of the symmetric positive semi-definite (or definite) matrix A_j corresponding to the smallest eigenvalue of A_j.

Figure 5.9 shows the INA of a 4-input 4-output system over the frequency range 0 → 1 rad/sec. Although the Gershgorin circles superimposed on the diagonal elements show that the basic system is diagonal dominant, the size of these circles at the 1 rad/sec end of the frequency range indicate that the interaction in the system may be unacceptable during transient changes. Using the pseudodiagonalisation algorithm described above, at a frequency of 0.9 rad/sec in each row, a simple wholly real compensator $\hat{K}_p$ can be determined which yields the INA shown in Figure 5.10. Here, we can see that the size of the Gershgorin discs has in fact been considerably reduced over all of the bandwidth of interest.

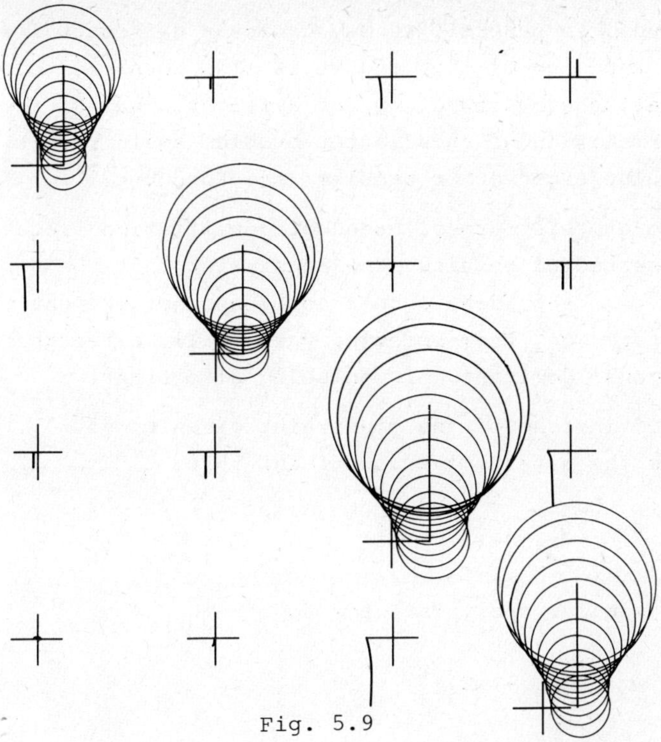

Fig. 5.9

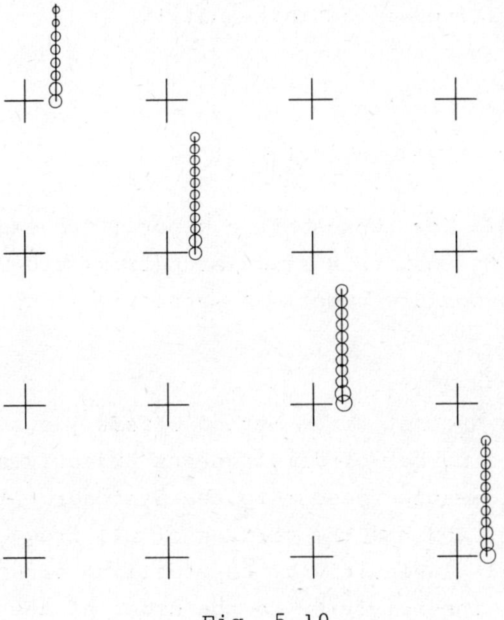

Fig. 5.10

However, in general, we may choose a different frequency ω for each row of $\hat{K}$, and it is also possible to pseudo-diagonalize each row of $\hat{Q}$ at a weighted sum of frequencies. The formulation of this latter problem again results in an eigenvalue/eigenvector problem (see Rosenbrock[1]).

Although this form of pseudodiagonalization frequently produces useful results, the constraint that the control vector $\hat{k}_j$ should have unit norm does not prevent the diagonal term $\hat{q}_{jj}$ from becoming very small, although the row is diagonal dominant, or vanishing altogether.

So, if instead of the constraint given by (5.33), we substitute the alternative constraint that

$$|\hat{q}_{jj}(j\omega)| = 1 \tag{5.41}$$

then a similar analysis leads to

$$A_j\hat{k}_j^T - \lambda E_j\hat{k}_j^T = 0 \tag{5.42}$$

where A_j is as defined by equation (5.37) and E_j is the symmetric positive semidefinite matrix

$$\begin{aligned} E_j &= \{e_{i\ell}^{(j)}\} \\ &= [\alpha_{ij}\alpha_{\ell j}+\beta_{ij}\beta_{\ell j}] \end{aligned} \tag{5.43}$$

Equation (5.42) now represents a generalized eigenvalue problem, since E_j can be a singular matrix, and must be solved using the appropriate numerical method.

5.5 Conclusions

The inverse Nyquist array method offers a systematic way of achieving a number of simultaneous objectives, while still leaving considerable freedom to the designer. It is easy to learn and use, and has the virtues of all frequency response methods, namely insensitivity to modelling errors including nonlinearity, insensitivity to the order of the system, and visual insight. All the results for inverse plots apply with suitable changes to direct Nyquist plots (see Reference 1, pp. 174-179). We can also obtain multivariable generalisations of the circle theorem for systems with nonlinear, time-dependent, sector-bounded gains[1,5]. Not only do the Gershgorin bands give a stability criterion; they also set bounds on the transfer function $h_{ii}(s)$ seen in the ith loop as the gains in the other loops are varied.

Several further ways of determining $\hat{K}_p(s)$ such that $\hat{Q}(s)$ is diagonal dominant are the subject of current research (see Leininger[6]). However, several industrial multivariable control problems have already been solved using this design method[7,8,9,10].

References

1. ROSENBROCK, H.H. : 'Computer Aided Control System Design', Academic Press, London, (1974).

2. MUNRO, N. and B.J. BOWLAND : 'The UMIST Computer-Aided Control System Design Suite', User's Guide, Control Systems Centre, UMIST, Manchester, England, 1980.

3. OSTROWSKI, A.M. : 'Note on Bounds for Determinants with Dominant Principal Diagonal', Proc. Am. Math. Soc. 3, (1952), pp. 26-30.

4. HAWKINS, D.J. : 'Pseudodiagonalisation and the Inverse Nyquist Array Methods', Proc. IEE, vol. 119, (1972), pp. 337-342.

5. COOK, P.A. : 'Modified Multivariable Circle Theorems', in Recent Mathematical Developments in Control (Ed. D.J. Bell), Academic Press, (1973), pp. 367-372.

6. LEININGER, G.G. : 'Diagonal Dominance for Multivariable Nyquist Array Methods Using Function Minimization', Automatica, vol. 15, (1979), pp. 339-345.

7. MUNRO, N. : 'Design of Controller for an Open-Loop Unstable Multivariable System Using the Inverse Nyquist Array', Proc. IEE, vol. 118, No. 9, (1972), pp. 1377-1382.

8. WINTERBONE, D.E., N. MUNRO and P.M.G. LOURTIE : 'Design of a Multivariable Controller for an Automotive Gas Turbine', ASME Gas Turbine Conference, Washington, Paper No. 73-GT-14, 1973.

9. MUNRO, N. and S. ENGELL : 'Regulator Design for the F100 Turbofan Engine', IEEE Int. Conf. on Control and its Applications, Warwick University, England, (1981), pp. 380-387.

10. KIDD, P.T., N. MUNRO and D.E. WINTERBONE : 'Multivariable Control of s Ship Propulsion System', Proc. of Sixth Ship Control Systems Symposium, Ottowa, Canada, October, 1981.

APPENDIX

To prove the result given by equation (5.21), we note that by (5.18a and b) and the conditions of the theorem there is no pole of $\hat{q}_{ii}$ on D, $i,j = 1,2,\ldots,m$, nor is there any zero of $\hat{q}_{ii}$ on D, $i = 1,2,\ldots,m$, because D is a compact set and $\hat{q}_{ii} > 0$ on D. Moreover, by Gershgorin's theorem, there is no zero of $|\hat{Q}|$ on D. Let $\hat{Q}(\alpha,s)$ be the matrix having

$$\begin{aligned} \hat{q}_{ii}(\alpha,s) &= \hat{q}_{ii}(s) \\ \hat{q}_{ij}(\alpha,s) &= \alpha\hat{q}_{ij}(s), \quad j \neq i \end{aligned} \tag{5.44}$$

where $\hat{q}_{ii}(s)$, $\hat{q}_{ij}(s)$ are the elements of $\hat{Q}(s)$ and $0 \leq \alpha \leq 1$. Then every element of $\hat{Q}(\alpha,s)$ is finite on D, and so therefore is $|\hat{Q}(\alpha,s)|$. Consider the function

$$\beta(\alpha,s) = \frac{|\hat{Q}(\alpha,s)|}{\prod_{i=1}^{m} \hat{q}_{ii}(s)} \tag{5.45}$$

which is finite for $0 \leq \alpha \leq 1$ and all s on D, and which satisfies $\beta(0,s) = 1$. Let the image of D under $\beta(1,\cdot)$ be Γ. Let the image of (0,1) under $\beta(\cdot,s)$ be γ_s. Then γ_s is a continuous curve joining the point $\beta(0,s) = 1$ to the point $\beta(1,s)$ on Γ. As s goes once round D, γ_s sweeps out a region in the complex plane and returns at last to its original position.

Suppose, contrary to what is to be proved, that Γ encircles the origin. Then the region swept out by γ_s as s goes once round D must include the origin. That is, there is some $\alpha(0,1)$ and some s on D for which $\beta(\alpha,s) = 0$. But the $\hat{q}_{ii}$ are all finite on D, so by (5.45) $|\hat{Q}(\alpha,s)|=0$.

By Gershgorin's theorem this is impossible. Hence the number of encirclements of the origin by Γ is, from (5.45),

$$0 = \hat{N}_Q - \sum_{i=1}^{m} \hat{N}_i \tag{5.46}$$

which is (5.21).

The Inverse Nyquist Array Design Method - Problems

P.1 A plant described by

$$G(s) = \begin{pmatrix} \frac{1}{s+1} & 0 \\ 0 & \frac{1}{s-1} \end{pmatrix}$$

has arisen from a system with

$$|T_G| = (s-1)(s+1)^2$$

If feedback is to be applied to this system through a feedback matrix $F = \mathrm{diag}\{1,2\}$, determine whether or not the resulting closed-loop system is stable.

Comment on cancellations occurring in the formulation of $G(s)$, and on the encirclements obtained in the resulting INAs.

{Acknowledgement for this problem and its solution are hereby made to Professor H.H. Rosenbrock of the Control Systems Centre, UMIST.}

P.2 Given a system described by the transfer-function matrix

$$G(s) = \begin{pmatrix} \frac{1}{s+1} & -\frac{1}{s+1} \\ \frac{1}{s+2} & \frac{1}{s+2} \end{pmatrix}$$

(i) State one reason why the inverse systems $G^{-1}(s)$ and $H^{-1}(s)$ are used in Rosenbrock's inverse Nyquist array design method.

(ii) Sketch the inverse Nyquist array for $G^{-1}(s)$, and comment on the diagonal dominance of the uncompensated system.

(iii) Determine a wholly real forward path compensator K such that the closed-loop system $H(s)$, with unity feedback, is decoupled.

(iv) Introduce integral action into both control loops and sketch the resulting root-locus diagrams for the final loop-tuning of the compensated system.

Chapter 6

Analysis and design of a nuclear boiler control scheme

Dr. F. M. HUGHES

Synopsis

The steam raising plant of a British designed nuclear power station for which the installed control scheme has given rise to performance and stability problems is considered. The lecture is based on studies carried out for the plant which illustrate the way in which multivariable frequency response methods can be used to analyse and identify the source of control problems and further enable alternative control schemes, having improved performance, to be designed.

6.1 Introduction

The role that multivariable frequency response methods can play in the analysis and design of control schemes for power plant is demonstrated. The particular plant dealt with comprises the drum boiler steam generating units of a nuclear power station for which the existing conventional feed control scheme has suffered from performance and stability problems. Although the boiler plant concerned is non-linear in nature, a great deal of useful information regarding dynamic behaviour can be obtained from the study of a linearised model. Under closed-loop control conditions, where the excursions in plant variables are held within tight bounds, a small variation, linearised model can give a realistic representation of controlled behaviour at a particular operating condition.

Using a suitable transfer function matrix model, the existing scheme is analysed and the reasons for its poor stability characteristics are quickly established via the multivariable frequency response approach. The design methods based on this approach are further used to devise an alternative control scheme possessing improved stability characteristics and performance capabilities. Although one particular power plant is dealt with, the analysis and design approaches adopted are applicable to power station drum boiler control in general.

6.2 Drum Boiler Plant and its Control Requirements

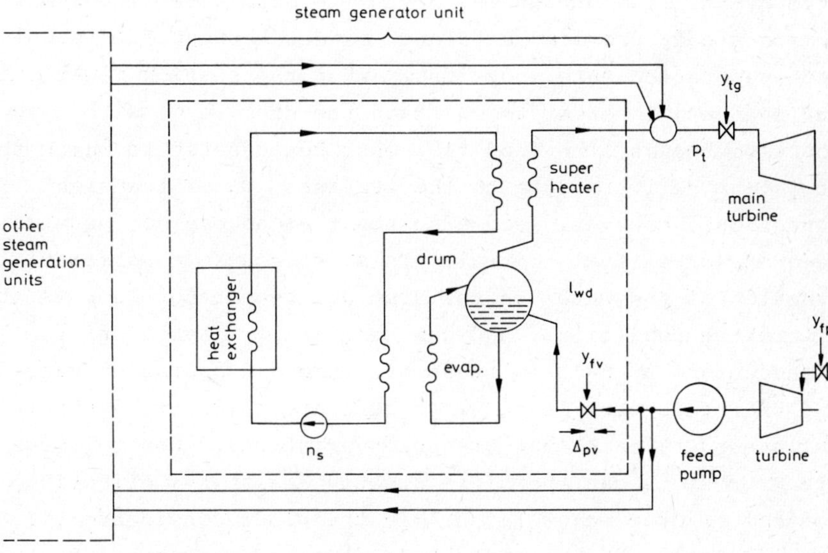

Fig. 6.1 Schematic Diagram of Plant.

A schematic block diagram of a nuclear station drum boiler is shown in Fig. 6.1 . The feedpump supplies water to the drum via the feedwater regulating valves. From the drum water enters the evaporator loop and returns as 'wet' steam to the drum. Then steam from the drum passes through the superheater, emerging as superheated (dry) steam for the main turbine supply. The heat energy input to the steam generators is obtained from the circulating coolant which transfers heat energy from the reactor. With the reactor temperature levels controlled the heat energy input can be varied by changing the coolant flow rate through adjustment of the coolant pump speed (n_s). Feed flow into the drum can be changed both by adjustment of the position of the feedwater regulating values (y_{fv}) and by adjustment of the steam input into the feedpump turbine (y_{fp}). The output of steam to the turbine, which determines the turbine generator power output, is controlled by adjustment of the position of the turbine governor valves (y_{tg}).

It is important to be able to control the level of water in the drum since disastrous consequences can result from the drum overfilling or running dry. The drum water level will remain steady provided a balance exists between feed supply and evaporation rate. Obviously when the station power output changes, in order to maintain the drum level within appropriate bounds, the feed flow must be adjusted to match the new steam delivery rate to the turbine. Under transient conditions, however, feed adjustment alone may not provide adequate drum level control. This is because a substantial fraction of the water in the drum and evaporator loop is at saturation conditions, and can be readily flashed off to steam during a fall in steam pressure. Evaporation rate and therefore drum level variations are therefore strongly influenced by transient pressure variations. Control over steam pressure can therefore greatly assist the effective control of drum water level. In the plant considered the main variables which need to be controlled are therefore the drum water level and the steam pressure[1].

Although not shown in Figure 6.1, the power plant comprises 3 steam generating units, each with its own drum, evaporator, superheater and separately controlled coolant flow. The feed to the drums is supplied by a common feedpump through individual feed control valves. Also, the three superheaters feed into a common steam header to provide a single supply to the main steam turbine.

6.3 Multivariable Frequency Response Methods

A linear multivariable control scheme can be represented most conveniently for frequency response analyses in terms of the block diagram of Figure 6.2 . The plant is modelled by the transfer function matrix $G(s)$ which relates its controllable inputs to its measurable outputs and the plant control scheme is represented by the precontroller matrix $K(s)$. The system open loop transfer function matrix is given as $Q(s) = G(s)K(s)$.

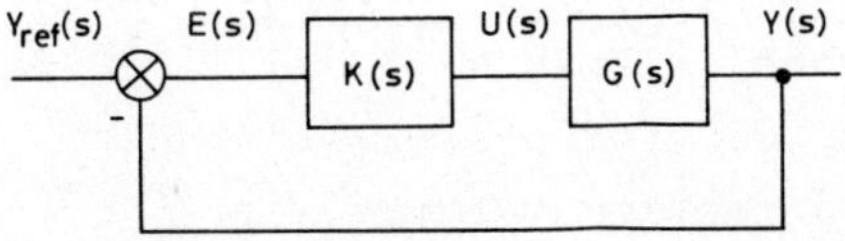

Fig. 6.2 Plant and controller representation.

If $G(s)$ happens to be diagonal then single loop stability assessment and design methods can be applied to each diagonal element in turn and the appropriate individual loop compensation needed to satisfy the specified performance thereby determined. In general, however, $G(s)$ will not be diagonal and the off diagonal elements which give rise to the interaction between the loops make analysis on an individual loop basis most unreliable. Alternative means are therefore required[2].

The system shown in Figure 6.2 has the closed-loop system transfer function $R(s)$ given by:

$$R(s) = (I+Q(s))^{-1}Q(s) = \frac{\mathrm{adj}(I+Q(s)).Q(s)}{\det(I+Q(s))} \qquad (6.1)$$

The poles of the closed-loop system, which determine stability, are the values of s for which $|R(s)| \to \infty$ and therefore the values of s for which $\det(I+Q(s)) = 0$.

If the $\det(I+Q(s))$ is mapped as s traverses the infinite radius semicircular contour in the right half s plane, then the Nyquist stability criterion can be applied to determine closed-loop stability from the number of encirclements that $\det(I+Q(s))$ makes about the origin. Although overall system stability is assessed information regarding the stability margins of individual loops is not provided.

An alternative approach to stability assessment is provided by re-expressing $Q(s)$ for any complex value of s, in terms of its eigenvalues and eigenvectors[3]:

$$Q(s) = U(s)\Lambda(s)U^{-1}(s) \tag{6.2}$$

where $\Lambda(s)$ is a diagonal matrix of characteristic transfer functions (diagonal elements $\lambda_i(s)$, $i = 1,2,\ldots,m$ for an m-input m-output control system) and $U(s)$ is a matrix of characteristic direction vectors.

It can readily be shown that:

$$\det(I+Q(s)) = \det(I+\Lambda(s)) \tag{6.3}$$

$$= (1+\lambda_1(s))(1+\lambda_2(s)(\ldots)(1+\lambda_m)(s)) \tag{6.4}$$

Hence overall stability can be assessed by applying the Nyquist stability criterion to each characteristic transfer function. If the net encirclements of the characteristic loci (i.e. the mappings of the characteristic transfer functions as s traverses the infinite semicircular contour) indicate stability then the overall closed loop system will be stable.

If the off-diagonal elements of $Q(s)$ are relatively small then the characteristic transfer functions are strongly related to the diagonal elements of $Q(s)$ and the individual loop transferences. The characteristic loci then give a useful indication of individual loop compensation requirements.

Since a desirable control scheme characteristic is usually

that it provides low loop interaction, an attempt to reduce interaction (i.e. the reduction of the off-diagonal elements of Q(s)) is normally the initial step in the design of the precontroller matrix K(s). Matrix K(s) performs column manipulations on G(s) and the manipulations required can be deduced from the frequency response array G(jω). Provided interaction has been made sufficiently small the individual control loop compensation requirements can be assessed and appropriate compensators designed.

However, in many systems it may prove very difficult to make all off diagonal elements of Q(s) (or $Q^{-1}(s)$ if the inverse Nyquist array method[4] is being used) small, and its attainment may result in a more complex control system than is strictly necessary for satisfactory control. Also, under some circumstances a certain level of interaction may be quite tolerable, so that all the off diagonal elements of Q(s) may not need to be small. The Q(s) matrix may well have large off diagonal elements, and the diagonal elements still approximate closely to the system characteristic transfer functions. This is obviously true when Q(s) is a triangular matrix. In such circumstances individual loop stability margins and compensation requirements can be assessed from the diagonal elements of Q(s) despite interaction levels being high for certain disturbances.

6.4 Analysis and Design Approach Adopted

In the drum boiler studies reported, the stability characteristics of the installed control scheme were assessed from its characteristic loci, and alternative schemes designed using the direct Nyquist array approach. The design aim for the alternative scheme was to determine a simple and effective control structure which provided good overall performance and improved stability characteristics. Simplicity of structure was ensured by restricting the precontroller for interaction reduction to a matrix of constants. Also, in the light of previous experience, rather than attempt to drastically reduce all off diagonal elements of Q(s), the function of the precontroller was made that of reducing enough off diagonal elements to ensure that the system characteristic loci were closely approximated by the diagonal elements of the resulting open loop frequency response array.

6.4.1 Transfer function model

The most convenient and flexible form of plant representation for use with multi-input multi-output frequency response methods is a matrix transfer function model, which relates controllable inputs and disturbances to measurable outputs. To simplify the plant representation for this analysis of fundamental dynamic characteristics and control requirements, station operation with symmetrical boiler conditions was considered. With all three steam generators having identical operating conditions, a single equivalent steam generator can be modelled. Using a comprehensive non-linear model of the plant the measurable system output responses were predicted for step changes in the measurable inputs.

The plant inputs considered were:

- (i) Coolant pump speed n_s
- (ii) Feedwater regulating value position y_{fv}
- (iii) Feedpump turbine steam input control y_{fp}
- (iv) Turbine governor valve position y_{tg}

The measurable outputs were:

- (i) Steam pressure upstream of turbine stop valves p_t
- (ii) Drum water level ℓ_{wd}
- (iii) Feedvalve pressure drop Δp_v

Having obtained step responses the equivalent plant transfer functions were derived using an interactive identification package with graphical display facilities. The package incorporated a model adjustment technique and minimised an integral of error squared criterion to match equivalent transfer function responses to the given plant response data.

The transfer function model obtained was of the following

form:

$$\begin{pmatrix} p_t \\ \ell_{wd} \\ \Delta p_v \end{pmatrix} = \begin{pmatrix} g_{11}(s) & g_{12}(s) & g_{13}(s) \\ g_{21}(s) & g_{22}(s) & g_{23}(s) \\ g_{31}(s) & g_{32}(s) & g_{33}(s) \end{pmatrix} \begin{pmatrix} n_s(s) \\ y_{fv}(s) \\ y_{fp}(s) \end{pmatrix} + \begin{pmatrix} g_{14}(s) \\ g_{24}(s) \\ g_{34}(s) \end{pmatrix} \begin{pmatrix} y_{tg} \end{pmatrix} \tag{6.5}$$

Variables n_s, y_{fv} and y_{fp} are the controllable inputs to the plant, whilst y_{tg} represents a disturbance input. The response of the plant following a disturbance in y_{tg} is representative of its performance following a change in station power output demand.

6.4.2 Analysis of existing control scheme

The existing control scheme, in terms of a single equivalent steam generating unit, is shown in Figure 6.3, and can be seen to consist of separate loops, (i.e. no interconnecting signals are employed). As a consequence of this and the loop philosophy adopted, the feed regulating valves had to be chosen to control drum level so that individual drum level control is provided under asymmetric operating conditions.

The actual steam raising plant consists of three drums each with its associated feedvalve for individual feed flow adjustment. However, if realistic valve openings are to be maintained, adjustment of feedvalve position alone has little effect on feed flow since an uncontrolled boiler feedpump acts as an almost constant flow device. In order to provide the substantial changes in feed flow needed for effective drum level control the feedvalve positional changes need to be backed up by changes in feedpump pressure. The required back up control is provided by a control loop which monitors changes in pressure drop across the feed valves and adjusts the feedpump to maintain at least a minimum differential pressure across each of the three feed regulator valves. Constant feed valve differential pressure makes feed flow approximately proportional to feed valve position.

Good control over drum level therefore requires a fast

acting feedpump control loop in addition to feedvalve adjustment. In the existing scheme the feedpump is controlled by adjusting the actuator which alters steam inlet flow to the feedpump turbine. This actuator is positioned by an open loop stepping motor which effectively introduces an integral term into the control loop.

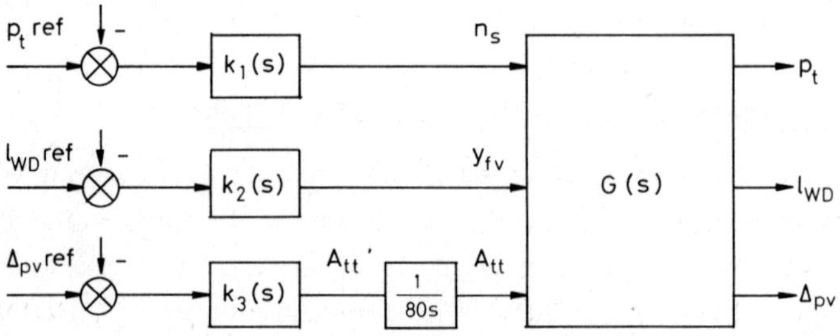

Fig. 6.3 Existing Feed Control Scheme.

The heat input to the steam generating unit can be modified by adjustment of the coolant flow via the coolant pump speed, and is used to ensure control over the steam pressure The open loop frequency response of the existing system, expressed in the form of a Nyquist Array is shown in Figure 6.4 . The individual loop compensators used were:

$$k_1(s) = .01[1 + \frac{1}{300} \cdot \frac{1}{s}] \tag{6.6}$$

$$k_2(s) = .13[1 + \frac{1}{300} \cdot \frac{1}{s}] \tag{6.7}$$

$$k_3(s) = .114 \tag{6.8}$$

It is immediately obvious from the array that the system is not diagonally dominant, and in particular diagonal element q_{22} is very small, being dwarfed by several of the off diagonal terms. This feature was to be expected since, as previously discussed, the alteration of feed valve position-alone has little effect on the boiler feed flow. The overall system stability characteristics cannot readily be assessed from an observation of this array; the off-diagonal elements are far too large for the diagonal elements to provide a useful guide to stability and the characteristic loci were calculated explicitly.

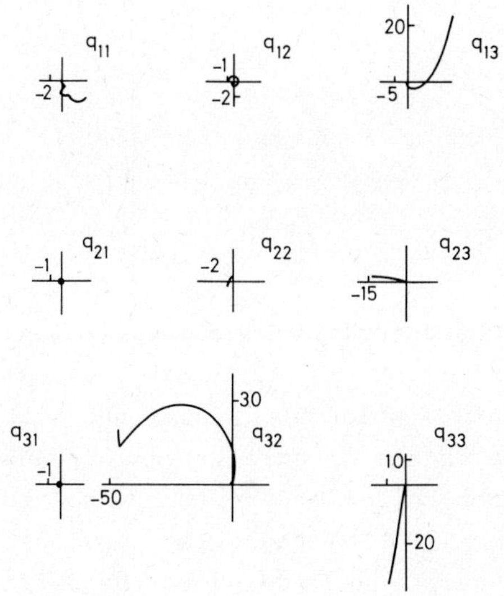

Fig.6.4 Nyquist Array of Existing Control Scheme.

The characteristic loci studies indicated that the parameter which had the most pronounced effect on system stability was the gain k_3. Two of the characteristic loci inferred wide stability margins and changed only slightly with variations to k_3. The third locus, however, changed drastically in both size and shape, and Figure 6.5 shows how system stability is impaired as k_3 is increased, with instability predicted at $k_3 = 0.4$. Instability of the oscillatory nature indicated has also been experienced on the plant itself. In addition the detrimental interaction between the loops which makes system stability sensitive to k_3 will also make stability and performance sensitive to changes in the operating conditions.

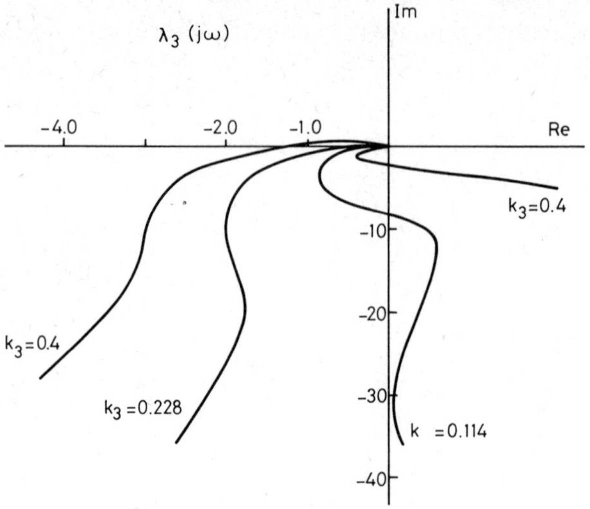

Fig. 6.5 Characteristic Locus Variation with Gain k_3.

A further feature which aggravates the stability of the existing scheme is the integral term introduced into the third loop by the boiler feedpump turbine actuator. This incurs an additional 90° of phase lag, obviously reducing the loop stability margin and adding to the difficulty of providing the stable, fast acting control loop needed to back up feed valve adjustments. This problem can be alleviated by introducing positional feedback into the actuator drive thus converting the actuator characteristic from $1/\tau s$ to $1/(1+\tau s)$. This lag transfer function characteristic provides a well defined, finite steady state gain and the faster the actuator drive is made, the lower is the effective time constant τ and the phase shift introduced.

6.4.3 Alternative control scheme design

As inferred in the previous section, the problems with the existing scheme are two-fold. The problem caused by the integral action characteristic of the stepper motor drive of the boiler feedpump turbine actuator can be alleviated by incorporating positional feedback. In the actuator position control loop so formed, a gain giving rise to a transfer function of 1/1+10s was considered realistic and used in the design studies. This change on its own, however, would still leave the problem of interaction between the loops unsolved, making it difficult to achieve the required response characteristics and stability margins over a wide operating range by means of individual loop compensation.

The design studies took as starting point the plant transfer function model with the modified feedpump turbine actuator characteristic, giving rise to the plant open loop frequency response array shown in Figure 6.6.

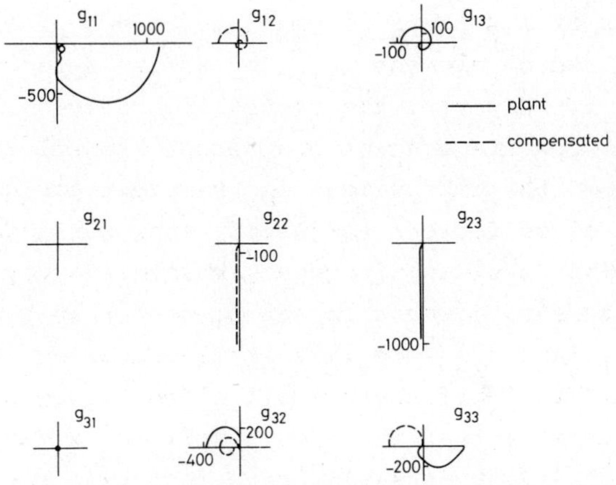

Fig. 6.6 Plant open loop frequency response array.

Various schemes were designed of varying complexity and performance characteristics[5], but only one particular scheme will be dealt with here. This scheme is considered to provide good stability and performance capabilities and possesses a simplicity which makes it readily extendible to the three drum version needed for implementation on the plant itself.

In designing an alternative control scheme the initial aim was to reduce interaction levels and the adopted criterion was that the plant input which most influenced a given output signal should be used for its control. In the open loop frequency response array of Figure 6.6, the largest element of row 2 is g_{23}, implying that drum level control is influenced most strongly by the boiler feedpump turbine (BFPT) actuator adjustment. Thus the most effective way of controlling drum level is via the feedpump turbine, rather than the feedwater regulator valves.

The required control configuration can be achieved by a precontroller matrix which swaps over column 2 with column 3, so that the large element g_{23} becomes a diagonal element in the open loop transfer function matrix.

The off diagonal q_{32} of the resulting array has a substantial steady state value, but this can be reduced by adding a factor of 1.2 times column 3 to column 2. It is also possible to reduce element q_{12} by adding a factor of column 1 to column 2. However, the reduction achieved is mainly effective at low frequencies and since integral control was to be employed the improvement in transient performance was not considered sufficient to justify the increased complexity involved. The resulting frequency response array is also shown in Figure 6.6, where it can be seen that the elements q_{21}, q_{31}, q_{13} and q_{23} are very small compared with the remaining elements. If these small elements are denoted by ε then the determinant of the return difference matrix which determines stability is given by

$$\det(I+Q(s)) = \begin{vmatrix} 1+q_{11} & q_{12} & \varepsilon \\ \varepsilon & 1+q_{32} & \varepsilon \\ \varepsilon & q_{32} & 1+q_{33} \end{vmatrix} \qquad (6.9)$$

i.e.

$$\det(I+Q(s)) \simeq (1+q_{11})(1+q_{22})(1+q_{33}) \quad \text{for } \varepsilon \text{ small} \qquad (6.10)$$

Although the open loop frequency response array does have substantial off diagonal elements in q_{12} and q_{32}, these have negligible effect on closed-loop stability. The system characteristic loci are approximated closely by the diagonal elements of $Q(j\omega)$ and individual loop stability characteristics and compensation requirements can be deduced from these.

The shape of the diagonal elements of the frequency response array indicated that the following individual loop controllers would provide good stability margins.

$$k_1(s) = 0.01(1 + \frac{1}{300}\frac{1}{s}) \quad (6.11)$$

$$k_2(s) = .05(1 + \frac{1}{300}\frac{1}{s}) \quad (6.12)$$

$$k_3(s) = .04(1 + \frac{1}{300}\frac{1}{s}) \quad (6.13)$$

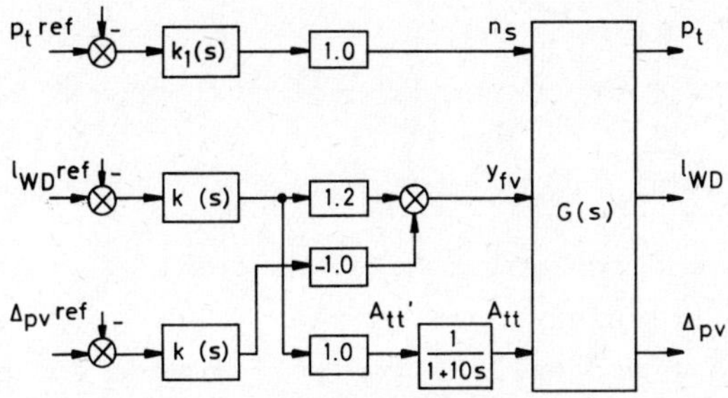

Fig. 6.7 Improved Control Scheme Configuration.

The resulting overall control scheme of Figure 6.7 has the characteristic loci shown in Figure 6.8, where it is obvious that much higher loop gains than those proposed could have been employed and reasonable stability margins still retained. On a linear system basis these higher gains would provide faster system responses. However, the system control actuators have physical limitations with respect to the amplitude and rate of change of output variations. Consequently, the relatively modest gains proposed were chosen to ensure that the demands made of the actuators are not unrealistically high.

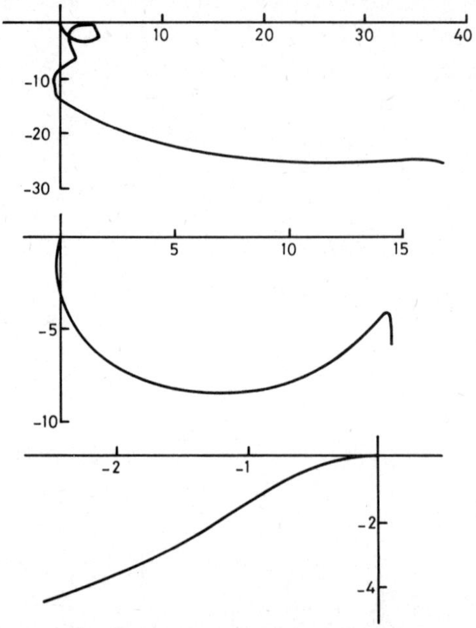

Fig. 6.8 Characteristic Loci of Improved Control Scheme.

6.5 Performance Assessment of Improved Scheme

The improved control scheme was tested by simulation using the linear transfer function model, and provided stable, adequately fast and well damped performance following changes in the reference levels of the control loops. A more realistic disturbance in practice is provided by a change in the position of the turbine governor control valves y_{tg}. Such disturbances occur during operation when the turbine governor gear responds to changes in the electricity system grid frequency. Responses of the designed scheme following a step change of 0.10 per unit (PU) in y_{tg} were computed, and the variation in steam pressure p_t, drum level l_{wd}, and differential pressure drop Δp_v are maintained well within acceptable bounds as indicated by the responses of Fig. 6.9. The non-minimum phase water level response due to steam flashing is well illustrated. The magnitudes of the actuator demands were examined and observed to be acceptable.

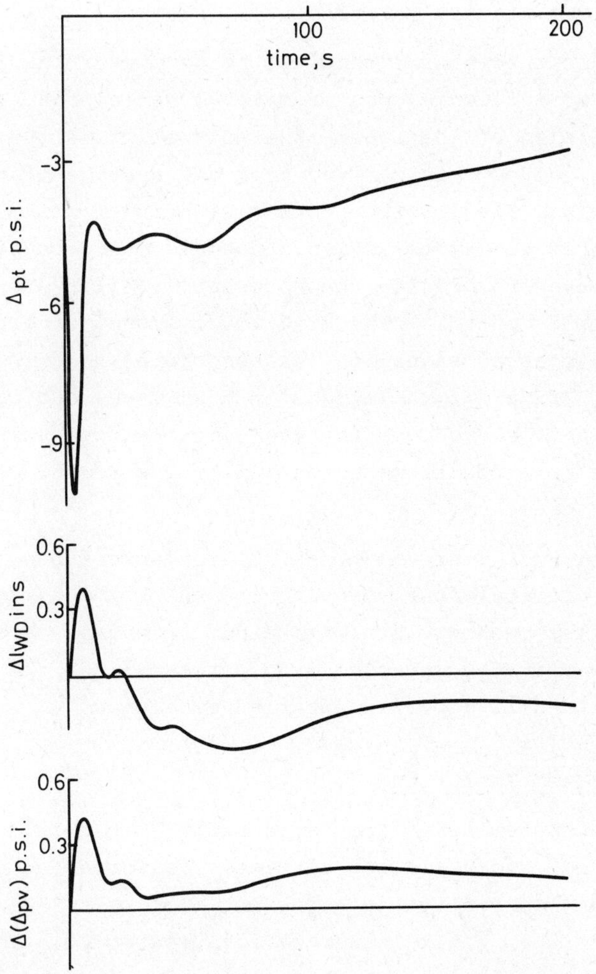

Fig. 6.9 System Response Following Step Change in Turbine Governor Valve position $y_{tg} = 0.1$ p.u.

6.5.1 Three drum version of improved scheme

In the design of the improved control scheme it was assumed that all three generating units possessed identical operating conditions enabling a single drum equivalent model to be used. Whilst this is reasonable for assessing fundamental dynamic characteristics and basic control requirements, any scheme devised must be capable of extension to the three drum case and provide good control in situations where drum operating conditions are not balanced[1].

The designed scheme used the feedpump turbine actuator as the major influence over drum level variations. In the three drum version of the scheme the control signal to the feedpump turbine actuator is derived from the average of the levels in the drums. The interconnecting signal (of gain .2) which was introduced from interaction and performance considerations also serves to provide the necessary facility for individual drum level control. The drum level signal of each drum provides a control signal to its associated feed regulating valve. Transient variations in drum level are controlled mainly by the feedpump turbine with the feed valves providing the required trimming action to ensure the correct level in each drum.

The control over the pressure drop across the feed valves is not critical, the main requirement being that reasonable feed valve openings are maintained. Control merely needs to be such that the pressure drop across all of the feed regulating valves is above a specified minimum.

6.6 System Integrity

For a system design to be acceptable it should be easy to commission and resilient to system failures. Specifically, it should remain stable and capable of being manually controlled, when control element failures occur. This aspect of system integrity, which is important during commissioning and in auto/manual changeover, can readily be checked in terms of the frequency response design approach.

System stability in the event of a controller channel not functioning can be assessed via the schematic block diagram of Figure 6.2 by introducing diagonal failure mode matrices between the controller and plant (F_a) and in the feedback path (F_f). F_a represents the actuator failure mode matrix and F_f the feedback failure mode matrix. The system open loop transfer function matrix Q_i is then given by

$$Q_i(s) \quad = \quad F_f G(s) F_a K(s) \tag{6.14}$$

Under normal operation with the control scheme fully functioning, $F_f = F_a = I$, so that

$$Q_i(s) \quad = \quad G(s)K(s) \quad = \quad Q(s) \tag{6.15}$$

If a controller element is disabled, the normal control signal is not transmitted and this situation can be represented by setting the appropriate diagonal element in the corresponding failure mode matrix to zero.

Commissioning situations where only certain control loops are closed can be represented by appropriate choice of F_f, and manual control situations where the operator adjusts the actuator rather than the normal control signal can be represented by appropriate choice of F_a. If, for example, the feedback channel of loop 1 is open for a 3-input 3-output system, then

$$F_f \quad = \quad \begin{bmatrix} 0 & 0 & 0 \\ 0 & 1 & 0 \\ 0 & 0 & 1 \end{bmatrix} \tag{6.16}$$

System stability is then assessed from either

$$\det(I+Q_i(s)) \quad = \quad \det(I+F_f G(s) F_a K(s)) \tag{6.17}$$

or from the characteristic loci of the open-loop system transfer function matrix $Q_i(s)$.

Such analysis has shown that the designed scheme can withstand all combinations of nonfunctioning controller elements without loss of stability. Consequently, any combination of automatic and manual control of loops will provide stable operation and complete freedom of choice over the order of loop closure is permitted when commissioning.

When a control system element fails during operation its associated output may change suddenly to its maximum or minimum value, resulting in a severe disturbance to the system. Wide and unacceptable variations in plant variables will then ensue unless appropriate signal freezing is carried out on detection of the fault. Such emergency action, however, obviously can only be effective if the resulting system is stable, and this fact can be determined from the integrity analysis approach outlined.

6.7 Conclusions

When analysing the existing control scheme the multivariable frequency response methods quickly enabled the sources of the control problems to be located and further provided the means for deriving an alternative scheme having improved stability and performance capabilities.

In the design of the alternative scheme, strict adherence to a particular multivariable frequency response method was not followed, neither was the normal objective of aiming for a low overall level of interaction. By taking advantage of the structural properties of the plant matrix model and the performance requirement of the plant, a simple control scheme having good stability characteristics was produced. Although an appreciable, but tolerable, level of interaction exists in the scheme, the system stability characteristics are dictated by the diagonal elements of the open loop transfer function matrix which can be used to determine individual loop stability margins and compensation requirements.

The simple structure of the control scheme, designed using a single drum equivalent plant model, enables it to be readily extended to the three drum version required for implementation on the plant itself.

The scheme designed also provides resilience to failures in that any combination of loops can be on automatic or manual control at any time, and complete freedom of choice over the loop closure sequence during commissioning is provided.

Attention has been confined to control scheme design and

analysis via a linear system model, but further simulation studies have been carried out on a comprehensive non-linear plant model. In these studies close agreement between linear and non-linear model performance predictions was observed, demonstrating that a linearised system representation can provide a good basis for the design of the basic structure of a control scheme.

References

1. COLLINS, G.B. and F.M. HUGHES : "The analysis and design of improved control systems for recirculation boilers", 8th IFAC World Congress, Kyoto, Japan, 1981.
2. MACFARLANE, A.G.J. : "A survey of some recent results in linear multivariable feedback theory", Automatica, vol. 8, No. 4, (1972), pp. 455-492.
3. BELLETRUTTI, J.J. and A.G.J. MACFARLANE : "Characteristic loci techniques in multivariable control system design", Proc. IEE, No. 118, (1971), pp. 1291-1297.
4. ROSENBROCK, H.H. : "Design of multivariable control systems using the inverse Nyquist array", Proc. IEE, vol. 117, (1969), pp. 1929-1936.
5. HUGHES, F.M. and G.B. COLLINS : "Application of multivariable frequency response methods to boiler control system design", Proc. 2nd International Conference on Boiler Dynamics and Control in Nuclear Power stations, pp. 411-418, British Nuclear Energy Society, London, 1980.

Acknowledgment

The contribution of the UKAEA (Winfrith) to the work described is gratefully acknowledged, particularly that of Mr G.B. Collins with whom much of the work was carried out.

Chapter 7

Optimal control

Dr. D. J. BELL

Synopsis

An overview of dynamic optimization is presented which commences with a statement of the important conditions from classical theory including the Weierstrass E-function and the second variation. The classical results are then reformulated in modern optimal control notation which leads immediately to a statement of Pontryagin's Minimum Principle. Necessary conditions for optimality of singular trajectories are deduced from the theory of the second variation. A description of Dynamic Programming is followed by a study of the Hamilton-Jacobi approach for the linear system/quadratic cost problem together with the associated matrix Riccati equation.

7.1 The Calculus of Variations: Classical Theory

The classical problem of Bolza is that of minimizing the functional

$$J = G(t_o, x(t_o), t_1, x(t_1)) + \int_{t_o}^{t_1} f(t,x,\dot{x})dt$$

(where x and $\dot{x}$ are n-vectors), subject to the differential constraints

$$\phi_\beta(t,x,\dot{x}) = 0 \qquad (\beta = 1,2,\ldots,k < n)$$

and end conditions

$$\psi_q(t_o, x(t_o), t_1, x(t_1)) = 0 \qquad (q = 1,2,\ldots, \leq 2n+2)$$

Defining a fundamental function $F = f + \lambda_\beta(t)\phi_\beta$ (note the double suffix summation is used throughout this chapter), where $\{\lambda_\beta\}$ is a sequence of time dependent Lagrange multipliers, a number of necessary conditions for a stationary value of J can be deduced from the first variation of J^1. These are (i) the Euler-Lagrange equations

$$\frac{d}{dt}\left(\frac{\partial F}{\partial \dot{x}_i}\right) - \frac{\partial F}{\partial x_i} = 0 \qquad (i = 1,2,\ldots,n) \tag{7.1}$$

(ii) the transversality conditions

$$\left(\left(F - \dot{x}_i \frac{\partial F}{\partial \dot{x}_i}\right)dt + \left(\frac{\partial F}{\partial \dot{x}_i}\right)dx_i\right)_{t_o}^{t_1} + dG = 0$$

$$d\psi_q = 0 \tag{7.2}$$

(iii) the Weierstrass-Erdmann corner conditions

$$\frac{\partial F}{\partial \dot{x}_i} \quad \text{and} \quad \left(F - \dot{x}_i \frac{\partial F}{\partial \dot{x}_i}\right) \qquad (i = 1,2,\ldots,n) \tag{7.3}$$

must be continuous across a corner of the extremal (i.e. the trajectory satisfying the necessary conditions).

(iv) the Weierstrass E-function condition, valid for all $t \in (t_o,t_1)$,

$$E = F(t,x,\dot{X},\lambda) - F(t,x,\dot{x},\lambda) - (\dot{X}_i-\dot{x}_i)\frac{\partial F}{\partial \dot{x}_i}(t,x,\dot{x},\lambda) \geq 0 \tag{7.4}$$

where a *strong* variation has been employed such that the derivative $\dot{X}$ on the variational curve may differ from the corresponding derivative $\dot{x}$ on the extremal by an arbitrary amount.

Note: If the fundamental function F does not contain the time t explicitly then a first integral of the Euler-Lagrange equations (7.1) exists and is

$$F - \dot{x}_i \frac{\partial F}{\partial \dot{x}_i} = \text{constant}, \; t \in (t_o,t_1)$$

The second variation of J is given by

$$d^2G + [(F_t - \dot{x}_i F_{\dot{x}_i})dt^2 + 2F_{\dot{x}_i} dx_i\, dt]_{t_o}^{t_1}$$

$$+ \int_{t_o}^{t_1} (F_{\dot{x}_i\dot{x}_k} \delta\dot{x}_i\delta\dot{x}_k + 2F_{\dot{x}_i x_k} \delta\dot{x}_i\delta x_k$$

$$+ F_{x_i x_k} \delta x_i \delta x_k)dt$$

which must be non-negative for all admissible variations $\delta x_i, \delta\dot{x}_i, dt_o, dt_1, dx_i(t_o)$ and $dx_i(t_1)$. This suggests a further minimization problem with the second variation as the functional and the variations δx_i and differentials dt_o, dt_1 as the unknowns, subject to the differential equations of constraint,

$$\frac{\partial\phi_\beta}{\partial\dot{x}_i}\delta\dot{x}_i + \frac{\partial\phi_\beta}{\partial x_i}\delta x_i = 0$$

Writing

$$\Omega = F_{\dot{x}_i\dot{x}_k} \delta\dot{x}_i\delta\dot{x}_k + 2F_{\dot{x}_i x_k} \delta\dot{x}_i\delta x_k + F_{x_i x_k} \delta x_i\delta x_k$$
$$+ \mu_\beta\left(\frac{\partial\phi_\beta}{\partial\dot{x}_i}\delta\dot{x}_i + \frac{\partial\phi_\beta}{\partial x_i}\delta x_i\right)$$

the variations δx which minimize the second variation must satisfy the Euler-Lagrange type equations

$$\frac{d}{dt}\left(\frac{\partial\Omega}{\partial\delta\dot{x}_i}\right) - \frac{\partial\Omega}{\partial\delta x_i} = 0$$

These equations are called the *accessory Euler-Lagrange equations* and the new minimization problem the *accessory minimum problem*.

By considering the second variation it can be shown[1] that a further necessary condition for optimality is that

$$F_{\dot{x}_i\dot{x}_k} \delta\dot{x}_i\delta\dot{x}_k \geq 0$$

subject to the constraints $\quad \dfrac{\partial\phi_\beta}{\partial\dot{x}_i}\delta\dot{x}_i = 0$

7.2 The Optimal Control Problem

The general optimal control problem may be formulated as follows. To find, in a class of control variables $u_j(t)$ $(j = 1,2,\ldots,m)$ and a class of state variables $x_i(t)$ $(i = 1,2,\ldots,n)$ which satisfy differential equations

$$\dot{x}_i = f_i(x,u,t)$$

and end conditions

$$x(t_o) = x_o , \quad t_o \text{ specified}$$

$$\psi_q(x(t_1),t_1) = 0 \quad (q = 1,2,\ldots, \leq n+1)$$

(t_1 may or may not be specified), that particular control vector $u = (u_1\ u_2,\ldots,u_m)^T$ and corresponding state vector $x = (x_1\ x_2,\ldots,x_n)^T$ which minimizes a cost functional

$$J = G(x(t_1),t_1) + \int_{t_o}^{t_1} L(x,u,t)dt$$

The control vector $u \in U$ where U is a control set which may be open or closed.

The essential differences between this problem and the classical one discussed earlier lies with the control vector $u(t)$. From practical considerations we must allow the control to be discontinuous. By raising the control vector u to the status of a derivative $\dot{z}$ we may treat z as an additional state vector and use the classical theory without further modification. A second difference lies in the fact that some or all of the control variables $u_j(t)$ arising in practical problems will be constrained by inequalities of the form

$$a_j \leq u_j \leq b_j$$

where a_j, b_j are usually known constants. To use the classical theory for a problem involving such control constraints the technique of Valentine can be used. This is to introduce additional control variables $\alpha_j(t)$ such that

$$(u_j - a_j)(b_j - u_j) - \alpha_j^2 = 0$$

These new equality constraints can then be treated in the same way as the set of system equations

$$\dot{x} = f = (f_1\ f_2, \ldots, f_n)^T$$

Let us apply the classical theory to the control problem formulated above. The fundamental function

$$F = L(x,u,t) + \lambda_i(t)\left[(f_i(x,u,t) - \dot{x}_i)\right] \tag{7.5}$$

$$(i = 1,2,\ldots,n)$$

With $\dot{z}_j = u_j$ the Euler-Lagrange equations are

$$\frac{d}{dt}\left(\frac{\partial F}{\partial \dot{x}_i}\right) = \frac{\partial F}{\partial x_i}, \quad \frac{\partial F}{\partial \dot{z}_j} = \text{constant } (j = 1,2,\ldots,m) \tag{7.6}$$

The transversality conditions are

$$\left(\left(F - \dot{x}_i \frac{\partial F}{\partial \dot{x}_i} - \dot{z}_j \frac{\partial F}{\partial \dot{z}_j}\right) dt + \frac{\partial F}{\partial \dot{x}_i}\, dx_i + \frac{\partial F}{\partial \dot{z}_j}\, dz_j\right)\Bigg|_{t_o}^{t_1} + dG$$

$$= 0 \tag{7.7}$$

$$d\psi_q = 0$$

The functions G and ψ_q do not involve the control variables explicitly. Since the variables z_j are not specified at either end point the transversality conditions yield

$$\frac{\partial F}{\partial \dot{z}j} = 0 \quad (j = 1,2,\ldots,m)$$

at t_o and t_1. Then, because of the second set of Euler-Lagrange equations (7.6), we have

$$\frac{\partial F}{\partial \dot{z}_j} \equiv \frac{\partial F}{\partial u_j} = 0 \quad t \in (t_o, t_1) \tag{7.8}$$

The remaining Euler-Lagrange equations (7.6) are equivalent

to

$$- \dot{\lambda}_i = \frac{\partial}{\partial x_i} (L + \lambda_i f_i) \tag{7.9}$$

Now define the Hamiltonian

$$H = L(x,u,t) + \lambda_i(t) f_i(x,u,t) \tag{7.10}$$

Equations (7.8)-(7.9) then take the form of the Pontryagin canonical equations

$$\frac{\partial H}{\partial u_j} = 0 \qquad (j = 1,2,\ldots,m) \tag{7.11}$$

$$\dot{\lambda} = - \frac{\partial H}{\partial x_i} \qquad (i = 1,2,\ldots,n)$$

Note that $F - \dot{x}_i \dfrac{\partial F}{\partial \dot{x}_i} = F + \lambda_i \dot{x}_i = F + \lambda_i f_i = H$

The transversality conditions (7.7) then reduce to

$$\left[H\, dt - \lambda_i dx_i \right]_{t_o}^{t_1} + dG = 0 \qquad d\psi_q = 0$$

where $\quad dG = \dfrac{\partial G}{\partial t_1} dt_1 + \dfrac{\partial G}{\partial x_i(t_1)} dx_i(t_1)$

and $\quad d\psi_q = \dfrac{\partial \psi_q}{\partial t_1} dt_1 + \dfrac{\partial \psi_q}{\partial x_i(t_1)} dx_i(t_1) \quad (q = 1,2,\ldots, \leq n+1)$

In particular, if $x(t_o)$ and t_o are specified but $x(t_1)$ and t_1 are free, then the transversality condition yields

$$H(t_1) = - \frac{\partial G}{\partial t_1}$$

$$\lambda_i(t_1) = \frac{\partial G}{\partial x_i(t_1)}$$

If, on the other hand, $x(t_o)$ and t_1 are specified and $\ell (<n)$ values specified at t_1 (with t_1 unspecified), e.g.

$$x_k(t_1) \quad (k = 1,2,\ldots,\ell) \text{ specified,}$$

then the transversality conditions yield

$$H(t_1) = -\frac{\partial G}{\partial t_1} \quad \text{and} \quad \lambda_s(t_1) = \frac{\partial G}{\partial x_s(t_1)}$$

$$(s = \ell+1, \ell+2, \ldots, n)$$

In the optimal control problem the Weierstrass-Erdmann corner conditions (7.3) are that H and λ_i $(i = 1,2,\ldots,n)$ must be continuous at points of discontinuity in $\dot{x}_i$ and u_j $(j = 1,2,\ldots,m)$.

The Weierstrass necessary condition (7.4) may now be rewritten as

$$E = L(x,\dot{Z},t) + \lambda_i(f_i(x,\dot{Z},t) - \dot{X}_i)$$

$$- L(x,\dot{z},t) - \lambda_i(f_i(x,\dot{z},t) - \dot{x}_i)$$

$$+ (\dot{X}_i - \dot{x}_i)\lambda_i - (\dot{Z}_j - \dot{z}_j)\frac{\partial}{\partial \dot{z}_j}(L + \lambda_i f_i - \lambda_i \dot{x}_i) \geq 0$$

On using equations (7.10)-(7.11), this inequality reduces to

$$L(x,\dot{Z},t) + \lambda_i f_i(x,\dot{Z},t) - L(x,\dot{z},t)$$

$$- \lambda_i f_i(x,\dot{z},t) \geq 0$$

Writing $\dot{Z}$, the control vector associated with a strong variation, as u and $\dot{z}$, the control vector associated with the extremal, as $\tilde{u}$ we are led to the inequality

$$H(x,\tilde{u},\lambda,t) \leq H(x,u,\lambda,t), \quad t \in (t_o,t_1)$$

Thus, the Hamiltonian H must assume its minimum value with respect to the control vector u at all points on the extremal. This is *Pontryagin's Minimum Principle.*

In our new notation, if the fundamental function F of (7.5) does not depend explicitly on the time t then a first integral of the Euler-Lagrange equations exists as mentioned in the Note above. This first integral is then

$$L + \lambda_i f_i - \lambda_i \dot{x}_i + \lambda_i \dot{x}_i = \text{constant}, \quad t \in (t_o,t_1)$$

i.e. $\quad H = \text{constant}, \quad t \in (t_o, t_1)$

Consider a one-parameter family of control arcs $u(t,\varepsilon)$ with the optimal control vector given by arc $u(t,o)$. The corresponding state vector x will also be a function of t and ε with $\varepsilon = 0$ along the optimal trajectory. Now introduce the notation for the state and control variations:

$$\eta_i = \left(\frac{\partial x_i}{\partial \varepsilon}\right)_{\varepsilon=0} \quad (i = 1,2,\ldots,n) \qquad \beta_j = \left(\frac{\partial u_j}{\partial \varepsilon}\right)_{\varepsilon=0} \quad (j = 1,2,\ldots,m) \tag{7.12}$$

and, in the case where t_1 is unspecified so that $t_1 = t_1(\varepsilon)$, the end-time variation

$$\xi_1 = \left(\frac{\partial t_1}{\partial \varepsilon}\right)_{\varepsilon=0} \tag{7.13}$$

The second variation must again be non-negative for a minimum value of J and is given by

$$\left((H_t - \dot{x}_i H_{x_i})\xi^2 + 2H_{x_i}\,\xi(\dot{x}_i\xi + \eta_i)\right)_{t=t_1} + d^2G$$

$$+ \int_{t_o}^{t_1} (H_{x_i x_k}\eta_i\eta_k + 2H_{x_i u_j}\eta_i\beta_j + H_{u_j u_h}\beta_j\beta_h)\,dt$$

subject to $\quad d^2\psi_q = 0$

$$\dot{\eta}_i = \frac{\partial f_i}{\partial x_k}\eta_k + \frac{\partial f_i}{\partial u_j}\beta_j$$

$$\frac{\partial \psi_q}{\partial t_1}\xi_1 + \frac{\partial \psi_q}{\partial x_{i1}}(\dot{x}_{i1}\xi_1 + \eta_{i1}) = 0$$

7.3 Singular Control Problems[3,4,5]

In the special case where the Hamiltonian H is linear in the control vector u it may be possible for the coefficient of this linear term to vanish identically over a finite interval of time. The Pontryagin Principle does not furnish any information on the control during such an interval and additional necessary conditions must be used to examine such extremals, known as singular extremals.

Definition

Let u_k be an element of the control vector u on the interval $(t_2,t_3) \subset (t_o,t_1)$ which appears linearly in the Hamiltonian. Let the (2q)th time derivative of H_{u_k} be the lowest order total derivative in which u_k appears explicitly with a coefficient which is not identically zero on (t_2,t_3). Then the integer q is called the order of the singular arc. The control variable u_k is referred to as a singular control.

Two necessary conditions will now be given for singular optimal control.

7.3.1 The generalized Legendre-Clebsch condition: a transformation approach

The generalized Legendre-Clebsch condition may be derived in an indirect way by first transforming the singular problem into a nonsingular one and then applying the conventional Legendre-Clebsch necessary condition. This condition for the transformed problem is the generalized Legendre-Clebsch condition for the original singular problem. We shall give Goh's approach[3] here which retains the full dimensionality of the original problem and is simpler in the case of vector controls. Kelley's transformation[5] reduces the dimension of the state space.

We have the second variation for a singular arc as

$$I_2 = \frac{1}{2}(\eta^T G_{xx}\eta)_{t_f} + \int_0^{t_f}(\frac{1}{2}\eta^T H_{xx}\eta + \beta^T H_{ux}\eta)dt \tag{7.14}$$

subject to the equations of variation

$$\dot{\eta} = f_x\eta + f_u\beta, \quad \eta(o) = 0 \tag{7.15}$$

Write $\beta = \dot{\zeta}$ and, without loss of generality, put $\zeta(o) = 0$.

Now write $\eta = \alpha + f_u \zeta$ in order to eliminate $\dot{\zeta}$ from the equations of variation. These equations become

$$\dot{\alpha} = f_x \alpha + (f_x f_u - \dot{f}_u)\zeta$$

The second variation becomes

$$I_2 = \left(\frac{1}{2}\alpha^T G_{xx} \alpha + \alpha^T G_{xx} f_u \zeta + \frac{1}{2}\zeta^T f_u^T G_{xx} f_u \zeta\right)_{t_f} + \int_0^{t_f} (\dot{\zeta}^T H_{ux}(\alpha + f_u \zeta) + \frac{1}{2}\alpha^T H_{xx}\alpha + \zeta^T f_u^T H_{xx}\alpha$$

$$+ \frac{1}{2}\zeta^T f_u^T H_{xx} f_u \zeta)dt$$

Integrate by parts the first term to eliminate $\dot{\zeta}$ from I_2. This gives

$$I_2 = \left(\frac{1}{2}\alpha^T G_{xx}\alpha + \alpha^T (G_{xx} f_u + H_{xu})\zeta + \frac{1}{2}\zeta^T (f_u^T G_{xx} f_u\right.$$

$$\left. + H_{ux} f_u)\zeta\right)_{t_f} + \int_0^{t_f} \left(\frac{1}{2}\zeta^T R^* \zeta + \zeta^T Q^* \alpha + \frac{1}{2}\alpha^T P^* \alpha\right)dt$$

We can now update ζ to the status of a derivative since $\dot{\zeta}$ has been eliminated completely from I_2 and the equations of variation. The conventional Legendre-Clebsch condition then yields $R^* \geq 0$.

Now, $R^* = f_u^T H_{xx} f_u - \frac{d}{dt} H_{ux} f_u - 2H_{ux}(f_x f_u - \dot{f}_u)$

$$= - (\ddot{H}_u)_u$$

Thus we arrive at the condition

$$(-)\frac{\partial}{\partial u}\left[\frac{d^2}{dt^2} H_u\right] \geq 0 \qquad t \in (t_o, t_f) \qquad (7.16)$$

This inequality is a generalized Legendre-Clebsch necessary condition in that it is similar in form to the classical Legendre-Clebsch necessary condition

$$\frac{\partial}{\partial u} H_u \geq 0 \qquad t \in (t_o, t_f)$$

If (7.16) is met with equality for all $t \in (t_o, t_f)$ then the functional I_2 is totally singular and another transformation must be made. In this way the generalized necessary conditions are obtained:

$$(-1)^q \frac{\partial}{\partial u} \frac{d^{2q}}{dt^{2q}} H_u \geq 0 \qquad t \in (t_o, t_f)$$
$$(q = 1, 2, \ldots .)$$

It can also be shown that

$$\frac{\partial}{\partial u} \frac{d^p}{dt^p} H_u = 0 \qquad t \in (t_o, t_f)$$
$$p \text{ an odd integer}$$

7.3.2 Jacobson's necessary condition[3]

First adjoin the equations of variation (7.15) to the expression for the second variation (7.14) by a Lagrange multiplier vector of the form $Q(t)\eta$, where Q is a $n \times n$ time-varying, symmetric matrix. Then

$$I_2 = \frac{1}{2} (\eta^T G_{xx} \eta)_{t_f} + \int_0^{t_f} (\tfrac{1}{2} \eta^T H_{xx} \eta + \eta^T Q (f_x \eta + f_u \beta - \dot{\eta})$$
$$+ \beta^T H_{ux} \eta) dt$$

Integrate by parts the last term of the integrand to give

$$I_2 = \frac{1}{2} (\eta^T (G_{xx} - Q) \eta)_{t_f} + \int_0^{t_f} (\tfrac{1}{2} \eta^T (\dot{Q} + H_{xx} + f_x^T Q + Q f_x) \eta$$
$$+ \beta^T (H_{ux} + f_u^T Q) \eta) dt$$

Now choose Q so that $\dot{Q} = -H_{xx} - f_x^T Q - Q f_x$

$$\text{with} \qquad Q(t_f) = G_{xx}(x(t_f), t_f) \tag{7.17}$$

$$\text{Then} \qquad I_2 = \int_0^{t_f} \beta^T (H_{ux} + f_u^T Q) \eta \, dt$$

with $\dot{\eta} = f_x\eta + f_u\beta \quad , \qquad \eta(o) = 0$

Now choose a special variation $\beta = \beta^*$, a constant for $t \in (t_1, t_1+\Delta T)$, $0 < t_1 < t_1 + \Delta T < t_f$, and $\beta = 0$ elsewhere. Then

$$I_2 = \int_0^{t_1+\Delta T} \beta^T (H_{ux} + f_u^T Q)\eta \, dt, \quad \eta(t_1) = 0 \quad (\eta \text{ continuous})$$

By the mean value theorem,

$$I_2 = \Delta T \, \beta^{*T}((H_{ux} + f_u^T Q)\eta)_{t_1+\theta\Delta T} \qquad (0 < \theta < 1)$$

Expanding in a Taylor's series:

$$I_2 = \Delta T \, \beta^{*T}((H_{ux}+f_u^T Q)_{t_1} \eta(t_1) + \theta\Delta T \frac{d}{dt}((H_{ux}+f_u^T Q)\eta)_{t_1} + 0(\Delta T^2))$$

which, since $\eta(t_1) = 0$, gives

$$I_2 = \Delta T \beta^{*T}(\theta\Delta T(H_{ux}+f_u^T Q)(f_x\eta + f_u\beta)_{t_1} + 0(\Delta T^2))$$

$$= \theta\Delta T^2 \beta^{*T}(H_{ux}+f_u^T Q)_{t_1} \; f_u(t_1)\beta^* + 0(\Delta T^3)$$

For an optimal solution, $I_2 \geq 0$. Thus,

$$(H_{ux} + f_u^T Q) f_u \geq 0 \qquad t \in (0, t_f) \qquad (7.18)$$

Equations (7.17) and inequality (7.18) constitute Jacobson's condition.

7.4 Dynamic Programming[6,7]

Consider the discrete system governed by the difference equations

$$x(i+1) = f^i(x(i), u(i)), \quad x(o) \text{ specified},$$
$$(i = 0,1,2,\ldots,N-1)$$

in which it is required to find the sequence of decisions (control vectors) $u(o), u(1), \ldots, u(N-1)$ which minimizes a cost function $\phi[x(N)]$.

The method of dynamic programming for solving such an optimal control problem is based on Bellman's *Principle of Optimality*. This states that an optimal policy has the property that whatever the initial state and initial decision are, the remaining decisions must constitute an optimal policy with regard to the state resulting from the first decision. In other words, having chosen $u(o)$ and determined $x(1)$, the remaining decisions, $u(1), u(2), \ldots, u(N-1)$ must be chosen so that $\phi[x(N)]$ is minimized for that particular $x(1)$. Similarly, having chosen $u(o), u(1), \ldots, u(N-2)$ and thus determined $x(N-1)$, the remaining decision $u(N-1)$ must be chosen so that $\phi[x(N)]$ is a minimum for that $x(N-1)$. Proof is by contradiction.

The principle of optimality leads immediately to an interesting computational algorithm. The basic idea is to start at the final end point $t = Nh$, where h is the step-length in time. Suppose that somehow we have determined $x(N-1)$. The choice of the final decision $u(N-1)$ simply involves the search over all values of $u(N-1)$ to minimize $\phi(x(N)) = \phi(f^{N-1}(x(N-1), u(N-1)))$.

Suppose we denote the minimum value of $\phi(x(N))$, reached from $x(N-1)$, by $S(x(N-1), (N-1)h)$, i.e.

$$S(x(N-1),(N-1)h) = \min_{u(N-1)} \phi(f^{N-1}(x(N-1),u(N-1)))$$

Similarly, we let

$$\begin{aligned} S(x(N-2),(N-2)h) &= \min_{u(N-2)} \min_{u(N-1)} \phi(x(N)) \\ &= \min_{u(N-2)} S(x(N-1),(N-1)h) \\ &= \min_{u(N-2)} S(f^{N-2}(x(N-2),u(N-2)),(N-1)h) \end{aligned}$$

In general, we are led to the recurrence relation

$$S(x(n),nh) = \min_{u(n)} S(f^n(x(n),u(n)),\ (n+1)h) \tag{7.19}$$

Using this recurrence relation and starting from $n = N-1$ the whole process is repeated stage by stage backwards to the initial point where $x(o)$ is specified. A sequence of optimal controls $u^o(N-1), u^o(N-2), \ldots, u^o(o)$ are thus generated. With the completion of this process it is possible to work forward in time from the specified point $x(o)$ since $u^o(o)$ will be known for that initial point. Work forward to $x(1), x(2), \ldots,$ using the optimal controls $u^o(o), u^o(1), \ldots,$ until the final end-point is reached with the final optimal control $u^o(N-1)$.

Clearly,

$$S(x(N),Nh) = \phi(x(N)) \tag{7.20}$$

and this is a boundary condition on S to be used in the solution of the difference equation (7.19) for S. In solving (7.19) the minimization process is carried out first and then, having found the optimal $u^o(n)$, S may be treated as a function depending explicitly upon $x(n)$. By its definition the optimum cost function must satisfy equation (7.19) and, furthermore, when S has been found then, by definition, we have found the optimum cost function. Therefore, the solution of (7.19) with boundary condition (7.20) is a necessary and sufficient condition for an optimum solution. Equation (7.19) is called the Hamilton-Jacobi-Bellman equation since it is closely related to the Hamilton-Jacobi equation of classical mechanics.

As the interval h is made smaller so the discrete problem usually goes over naturally into the continuous problem. The general form of the recurrence relation (7.19), to first order terms in h, then gives rise to the equation

$$S(x,t) = \min_u \left(S(x,t) + \sum_i \frac{\partial S}{\partial x_i} h f_i(x,u) + h \frac{\partial S}{\partial t} \right)$$

Since S is not a function of u this last equation reduces to

$$\frac{\partial S}{\partial t} + \min_u \left[\frac{\partial S}{\partial x_i} f_i(x,u)\right] = 0 \tag{7.21}$$

u must be chosen to minimize the sum $f_i \frac{\partial S}{\partial x_i}$ resulting in a control law $u(x,t)$. The partial differential equation (7.21) may then be solved for $S(x,t)$ subject to the boundary condition

$$S(x(t_1),t_1) = \phi(x(t_1)) \tag{7.22}$$

7.5 The Hamilton-Jacobi Approach[2,7]

This approach to the problem of optimal control is an alternative to Pontryagin's Minimum Principle. It has two advantages: (i) it eliminates the difficult two-point boundary value problem associated with the Minimum Principle, (ii) it yields a closed-loop solution in the form of an optimal control law $u^o(x,t)$. A disadvantage is that it gives rise to a nonlinear partial differential equation, the Hamilton-Jacobi equation, and a closed form solution has been obtained for only a few special cases. This difficulty arises since only one set of initial conditions was considered in the Pontryagin approach whereas here, by stipulating a closed-loop solution, we are seeking the solution for a whole class of initial conditions.

Let the function $S(x,t)$ of (7.21) be given as the optimal cost function

$$S(x,t) = \int_t^{t_1} L(x(\tau),u^o(x,\tau),\tau)d\tau \tag{7.23}$$

i.e. $S(x,t)$ is the value of the cost function evaluated along an optimal trajectory which begins at a general time t and associated state $x(t)$. It can be shown that the partial derivatives $\partial S/\partial x_i$ of equation (7.21) have a time behaviour identical with the Lagrange multipliers λ_i in the Pontryagin approach. Thus, with the cost function given in (7.23), equation (7.21) may be written as

$$H(x, \nabla S(x,t),t) + \partial S/\partial t = 0 \tag{7.24}$$

This equation is known as the Hamilton-Jacobi equation. The boundary condition (7.22) becomes

$$S(x(t_1),t_1) = 0$$

In the case of a linear, time-invariant system

$$\dot{x} = Ax + Bu$$

with a quadratic cost function

$$\int_0^{t_1} (x^TQx + u^TRu)dt$$

(Q and R constant, symmetric matrices with R positive definite) the solution of the Hamilton-Jacobi equation can be obtained in closed form. The Hamiltonian is

$$H = x^TQx + u^TRu + \nabla S^T(Ax+Bu)$$

Assuming u is unbounded,

$$\frac{\partial H}{\partial u} = 2Ru + B^T \nabla S = 0$$

whence $$u^o = -\frac{1}{2} R^{-1} B^T \nabla S(x,t) \qquad (7.25)$$

Substituting this result into equation (7.24) yields the Hamilton-Jacobi equation

$$(\nabla S)^T Ax - \frac{1}{4}(\nabla S)^T BR^{-1} B^T \nabla S + x^TQx + \frac{\partial S}{\partial t} = 0$$

It can be shown that $S(x,t) = x^TP(t)x$ $(P > 0)$ is a solution to this equation provided that the matrix P satisfies the *Matrix Riccati equation*

$$\dot{P} + Q - PBR^{-1}B^TP + PA + A^TP = 0 \qquad (7.26)$$

with boundary condition

$$P(t_1) = 0 \qquad (7.27)$$

A relatively straightforward method for solving (7.26) subject to (7.27) is a backward integration on a digital computer. Once the matrix P is known the optimal control law is found from (7.25) as

$$u^o(x,t) = -R^{-1}B^TPx$$

$$= -K^T(t)x \quad \text{say.}$$

The elements of the matrix $K(t)$ are known as the *feedback coefficients*.

In the case where the final time is infinite the matrix P is constant. We then only have to solve a set of algebraic equations ($\dot{P} = 0$ in (7.26))

$$Q - PBR^{-1}B^TP + PA + A^TP = 0$$

the *reduced* or *degenerate Riccati equation*. Of the several possible solutions to this nonlinear equation in the elements of the matrix P the one required is that yielding a positive definite P.

References

1. BLISS, G.A. : Lectures on the Calculus of Variations. University of Chicago Press, 1961.
2. BRYSON, A.E. and Y.C. HO : Applied Optimal Control, Blaisdell, 2nd edition, 1975.
3. BELL, D.J. and D.H. JACOBSON : Singular Optimal Control Problems, Academic Press, 1975.
4. CLEMENTS, D.J. and ANDERSON, B.D.O. : Singular Optimal Control : The Linear-Quadratic Problem. Springer-Verlag, 1978.
5. KELLEY, H.J., R.E. KOPP and H.G. MOYER : Singular Extremals in "Topics in Optimization" (G. Leitmann, ed.) Academic Press, 1967.
6. BURLEY, D.M. : Studies in Optimization. Intertext Books 1974.
7. DENN, M.M. : Optimization by Variational Methods, McGraw-Hill, 1969.

Problems

P.7.1 Apply Pontryagin's principle to the problem of minimising the time taken for a system to transfer from the given state $(-1 \;\; 2)^T$ at time zero to the origin at time t_1 when the system equations are

$$\dot{x}_1 = x_2, \quad \dot{x}_2 = -x_1 + u \quad (|u| \leq 1)$$

Show that the phase trajectories are circles and that the phase point makes a complete revolution in time 2π and in a clockwise direction. Describe the phase trajectory to the origin assuming there is only one switching point for the control u. Explain why the total time for the phase trajectory must be less than $3\pi/2$.

P.7.2 A system is governed by the equations

$$\dot{x}_1 = \cos x_3, \quad x_1(0) = 0, \quad x_1(t_f) = \sqrt{2}$$

$$\dot{x}_2 = \sin x_3, \quad x_2(0) = 0, \quad x_2(t_f) = 1$$

$$\dot{x}_3 = u \quad , \quad x_3(0) = 0, \quad x_3(t_f) \text{ free}, |u| \leq 1$$

It is required to minimise the final time t_f . Show that there is a partially singular control commencing with u = 1 $t \in (0, \pi/4)$ which yields a candidate trajectory satisfying all the constraints. Verify that the GLC condition is satisfied along the singular subarc.

Chapter 8

Control system design via mathematical programming

Professor D. Q. MAYNE

Synopsis

Control design requires the choice both of strúcture of the controller and parameters for a given structure; the first task is guided by system theory; the latter is aided by mathematical programming. For a given control structure design objectives can often be specified as satisfying a set of inequalities or minimizing a function subject to these inequalities. It is shown that many objectives can be expressed as infinite dimensional constraints ($\phi(z,\alpha) \leq 0$ for all $\alpha \in A$). Algorithms for the resultant semi-infinite programming problems are briefly outlined.

8.1 Introduction

Control design is a fascinating mixture of control theory, numerical techniques and the experience and intuition of the designer. Control theory yields important structural results such as the structure of a controller necessary to achieve asymptotic tracking and regulation for all desired outputs (reference signals) and disturbances lying in specified classes; it also yields synthesis techniques, that is, solutions to well-defined problems such as optimal control (linear system, quadratic cost case) and pole allocation. The problems solved in this way are often too simple (being well defined) to represent practical design situations yet the resulting synthesis techniques may often be useful in solving recurring sub-problems in the design process. Given the structure of the controller, parameters have to be chosen. Theoretical results are often of little help at this stage. For example, multivariable root locus theory may yield the asymptotic root loci (and aid the choice of a controller structure) but not suitable values for (finite) controller parameters; a numerical procedure (e.g. in the form of a root locus plot) is necessary. The complexity of most design problems requires the constant interaction of a designer, specifying and changing some of the many objectives and constraints, choosing suitable structures and intervening where necessary in the many complex calculations.

Given the active intervention of the designer, it appears

that many, if not most, of the design specifications can be expressed as inequalities. It is, of course, possible that some *a priori* constraints are unrealistic and must therefore be altered during the design process; a design may also proceed by initially satisfying easily attained constraints and then tightening these constraints. A computer aided design package can provide these interactive facilities.

Many design specifications can be expressed as conventional inequalities. Thus the requirement of stability of a linear multivariable system can be expressed as $R\lambda^i(z) \leq 0$, $i = 1,\ldots,n$ where $\lambda^i(z)$ denotes the ith eigenvalue of the closed loop system given that the controller parameter is z. Nevertheless many design constraints are considerably more complex and have the form $\phi(z,\alpha) \leq 0$ for all $\alpha \in A$ where A is an interval or, even, a subset of R^m, $m > 1$. These *infinite dimensional* constraints have only recently been considered in the mathematical programming literature; nevertheless algorithms for satisfying such constraints or for minimizing a function subject to these constraints are now available and provide useful new design aids.

It is interesting to note that infinite dimensional inequalities arise in a wide variety of applications. One example is the design of a filter such that the output ψ to a given input pulse s satisfies an envelope constraint ($\psi(t) \in [a(t), b(t)]$ for all $t \in [0, t_1]$). This problem is relevant to pulse compression in radar systems, waveform equalization, channel equalization for communications and deconvolution of seismic and medical ultrasonic data. Another example is the design of framed structures such that the displacement of the structural elements in response to a specified disturbance is limited at every time in a given interval; the differential equations describing the motion of the structure are non-linear. In these two examples A is an interval. Consider, however, a circuit design problem where the parameters of the manufactured circuit differ from the (nominal) design value $z \in R^r$ by a tolerance error $t \in R^r$ known to lie in a tolerance set $A \subset R^r$. If the constraints must be met by all possible manufactured circuits then constraints of the form $\phi(z+t) \leq 0$ for all $t \in A$ must be satisfied. A similar problem occurs, for example, in the design of chemical plant with uncertain parameters. If the uncertainty or tolerance error occurs in parameters other than the design parameter then the constraint becomes $\phi(z,\alpha) \leq 0$ for all $\alpha \in A$, A a subset of R^m say. This can be expressed as $\max\{\phi(z,\alpha) | \alpha \in A\} \leq 0$.

Even more complex constraints can occur. In many circumstances post manufacture tuning (or trimming) of the design parameter z is permitted. If $z + r(q)$ denotes the value of the tuned parameter, then a design z is acceptable if for all $\alpha \in A$ there exists a $q \in Q$ such that $\phi(z+r(q),\alpha) \leq 0$. This can be expressed as:

$$\max_{\alpha\in A} \min_{q\in Q} \phi(z+r(q),\alpha) \leq 0$$

Further complexities arise when the properties of the function ϕ are considered. In many cases ϕ is a differentiable function (of z and α); in other cases, however, ϕ is not. An important example of this is discussed in the next section; many control design objectives can be specified as constraints on singular values of transfer function matrices and singular values are not differentiable functions.

This paper shows how infinite dimensional constraints arise naturally when design objectives for linear and non-linear control systems are formulated. The design of linear and non-linear multivariable systems is discussed in Sections 8.2 and 8.3 respectively. Algorithms for solving the resulting feasibility and optimization problems are discussed in Section 8.4 .

8.2 Linear Multivariable Systems

Design objectives for linear multivariable systems are discussed in Reference 1 where several current approaches to control system design are presented. If $Q(z,s)$ denotes the loop transfer function when the controller parameter is z

and $T(z,s) \triangleq I + Q(z,s)$ the matrix return difference, then $\det[T(z,s)] = \rho_c(z,s)/\rho_0(z,s)$ where $\rho_0(z,s)$ and $\rho_c(z,s)$ are the open and closed loop characteristic polynomials so that stability can be assessed in the usual way from the Nyquist plot of $\theta(z,\omega) \triangleq \det[T(z,j\omega)]$. Thus, if the system is open loop stable, the closed loop system is stable if $\phi(z,\omega) \leq 0$ for all $\omega \in [0,\infty)$ where $\phi(z,\omega) \triangleq [I\theta(z,\omega)] - \alpha[R\theta(z,\omega)]^2 + \beta$ (R and I denote the "real part" and "imaginary part" respectively) and α and β are chosen so that the origin does not lie in the set $\{s \in C \mid s = u + jv, v - \alpha u^2 + \beta \leq 0\}$ (a suitable choice is $\alpha = 1$, $\beta = 0.2$). In practice, due to the fact that $Q(z,j\omega) \to 0$ as $\omega \to \infty$, the semi infinite set $[0,\omega)$ can be replaced by $[0,\omega_1]$ so that the stability constraint can be expressed as the infinite dimensional constraint:

$$\phi(z,\omega) \leq 0 \quad \text{for all } \omega \in [0,\omega_1] \qquad (8.1)$$

Although a conventional inequality ($\phi(z) \leq 0$ where $\phi(z) \triangleq \max\{R\lambda^i(z) \mid i = 1,\ldots,n\} - \varepsilon$, $\lambda^i(z)$ being the ith eigenvalue of the closed system) may be used to ensure stability when the system is finite dimensional, constraint (8.1) may be employed for infinite systems.

Performance (e.g. small tracking error) can be expressed as an upper bound on $||T^{-1}(z,j\omega)||_2$ for all $\omega \in [0,\omega_1]$ say (T^{-1} is the transfer function from the reference to the error). This can be expressed as in (8.1) if $\phi(z,\omega) \triangleq a(\omega) - \underline{\sigma}(z,\omega)$ where $\underline{\sigma}(z,\omega)$ is the *smallest* singular value of $T(z,j\omega)$ and $a(\omega)$ is a (large) positive function of ω; the constraint $\phi(z,\omega) \leq 0$ is equivalent to $\underline{\sigma}(z,\omega) \geq a(\omega)$, i.e. to a large loop gain. On the other hand, the transfer function from the sensor noise to the output is

$$Q(z,s)T^{-1}(z,s) = T^{-1}(z,s)Q(z,s) = [I+Q^{-1}(z,s)]^{-1} ;$$

a constraint on the magnitude of the component of the output due to sensor noise can be expressed as $\bar{\sigma}[I+Q^{-1}(z,j\omega)^{-1}] \leq b(\omega)$ or $\underline{\sigma}[I+Q^{-1}(z,j\omega)] \geq c(\omega)$ for all ω. This can also be expressed in the form of (8.1) . The performance constraint requires $||T^{-1}||$ to be small and the output noise constraint requires $||T^{-1}Q||$ to be small; since $T^{-1} + T^{-1}Q = [I+Q]^{-1}[I+Q]. = I$, these constraints conflict[2]. Similar expressions can be employed to impose constraints on the control action in response to reference signals, disturbances and sensor noise[2].

While existing frequency response design methods can be employed to ensure integrity (maintenance of stability) in the face of large variations in loop gains (due e.g. to actuator or sensor failure) they cannot necessarily cope with large variations in plant parameters. The reason for this is that it is possible[2] for small variations in the plant to cause large variations in $\det[T(z,s)]$ (or in the characteristic loci), so that even though the Nyquist plot of $\det[T(z,j\omega)]$ has large gain and phase margins, the result design is not necessarily robust. It is argued[2] that the percentage variation in the model of the plant $(||\Delta G(j\omega)G^{-1}(j\omega)||)$ is likely to be higher at high frequencies. It is also shown that stability can be ensured for all plant variations satisfying $\bar{\sigma}[\Delta G(j\omega)G^{-1}(j\omega)] \leq \ell(\omega)$ if $\underline{\sigma}[Q(z,j\omega)T^{-1}(z,j\omega)] < 1/\ell(\omega)$ for all $\omega \in [0,\infty)$; $\ell(\omega)$ commonly increases with ω and may exceed unity. This stability constraint may also be expressed in the form of (8.1) and also conflicts with the performance constraint; in practice a pass band must be chosen so the performance (and stability) constraints are satisfied for ω in this band while the stability constraints dominate outside (where $1/\ell(\omega)$ is small).

The variations in the plant transfer function $G(s)$ considered above are completely unstructured (all variations $\Delta G(s)$ satisfying $\bar{\sigma}[\Delta G(j\omega)G^{-1}(j\omega)] \leq \ell(\omega)$ are permitted). If more precise information is available (e.g. certain plant parameters p always lie in a given set P) then the robustness can be ensured by the modified stability constraint

$$\phi(z,p,\omega) \leq 0 \quad \text{for all } \omega \in [0,\omega_1], \text{ all } p \in P \qquad (8.2)$$

where

$$\phi(z,p,\omega) \triangleq [I\theta(z,p,\omega)] - \alpha[R\theta(z,p,\omega)]^2 + \beta$$

and

$$\theta(z,p,\omega) \triangleq \det[T(z,p,j\omega)]$$

the return difference being expressed as a function of the controller parameter z, the plant parameter p and the complex frequency s. If q denotes certain plant or controller parameters that can be tuned (after the controller has been implemented) and Q the tuning range then constraint (8.2) is replaced by:

for all $\omega \in [0,\omega_1]$, all $p \in P$ there exists a $q \in Q$ such that $\phi(z,p,q,\omega) \leq 0$

or, equivalently:

$$\max_{\omega\in\Omega,p\in P} \min_{q\in Q} \phi(z,p,q,\omega) \leq 0$$

where $\Omega \triangleq [0,\omega_1]$ and ϕ is defined as above with $T(z,p,s)$ replaced by $T(z,p,s)$.

It is, of course, easy to express constraints on time responses of outputs, states, controls or control rates in the same form[3]. Thus constraints in overshoot, rise time and settling time on the unit step response of an output y may be expressed as:

$$\left.\begin{aligned} y(z,t) &\leq 1.15 \quad \text{for all} \quad t \in [0,t_2] \\ y(z,t) &\geq 0.5 \quad \text{for all} \quad t \in [t_1,t_2] \\ |y(z,t) - 1| &\leq 0.02 \quad \text{for all} \quad t \in [t_2,t_3] \end{aligned}\right\} \qquad (8.3)$$

If robustness to plant variations is required then $y(z,t)$ should be replaced by $y(z,p,t)$ and constraints (8.3) be required to hold for all $p \in P$.

It is thus apparent that many, if not all, of the design objectives may be expressed as infinite dimensional constraints. How can these objectives be achieved? If compenpensation is employed to yield a diagonally dominant transfer function the controllers for each loop can be independently designed to satisfy many design objectives (but not necessarily robustness to plant variations). The characteristic loci, on the other hand, do not correspond to separate loops; however, the associated design method employs an alignment technique so that (at least at selected frequencies) the loci appear to the designer to be associated with distinct loops. Again, many design objectives (but not necessarily robustness with respect to plant variation) can be achieved with this method.

Since pre-compensation cannot be used to isolate, for example, singular values (and would not in any case be desirable) more sophisticated adjustment techniques, such as those provided by mathematical programming, are required to choose suitable parameters.

8.3 Non-linear Systems

Much less effort has been devoted to the problem of designing non-linear multivariable controllers because of the overwhelming complexity of the task. Of course frequency response criteria for systems of special structure (e.g. linear dynamics systems in tandem with static non-linearities) and heuristic criteria, (e.g. describing functions for assessing stability) have been employed for design. Lyapunov methods can be employed for systems whose state dimension does not exceed three or four. In practice, simulations of the non-linear are often employed to assess stability and performance. Theoretically this requires simulations over the semi-infinite interval $[0,\infty)$ for every initial state of interest. We show in this section how stability and performance constraints can be expressed as infinite dimensional inequalities whose evaluation is considerably less expensive computationally than simple simulation.

Suppose the controlled system is described by:

$$\dot{x}(t) = f(x(t),z) \tag{8.4}$$

where, as before, z denotes the controller parameter, and f is continuously differentiable. Suppose that the set of initial states of interest is a compact subset Z of R^n. Let $V : R^n \to R$ be a continuous function with the following properties:

(i) $V(x) \geq 0$ for all $x \in R^n$,

(ii) $V(\alpha x) = \alpha V(x)$ for all $\alpha \in [0,\infty)$ and all $x \in R^n$,

(iii) $V(x) = 0$ if and only if $x = 0$.

An example of such a function is $x \to (x^T P x)^{\frac{1}{2}}$ where P is positive definite. For all x in R^n let the set $B(x)$ be defined by:

$$B(x) \triangleq \{x' \in R^n | V(x') \leq V(x)\}$$

Clearly $B(0) = \{0\}$, $B(\alpha x) = \alpha B(x)$, $B(\alpha x) \subset B(x)$ if $\alpha < 1$ and $B(x') \subset B(x)$ if and only if $V(x') \leq V(x)$. Choose $\bar{x}$ so that $x \subset \bar{B} \triangleq B(\bar{x})$; let $\bar{V} \triangleq V(\bar{x})$. Let $x(t,z,x_0)$ denote the solution of equation (8.4) with $x(0) = x_0$. If z can be chosen so that $\dot{V}(x,z) \triangleq V_x(x)f(x,z) < 0$ for all $x \in \bar{B}$, $x \neq 0$, then the closed loop system is asymptotically stable with a region of attraction $\bar{B}$ ($V(\cdot,z)$ is a Lyapunov function). This is seldom possible; however, it is possible to relax the requirement $\dot{V}(x,z) < 0$ for all $x \in \bar{B}$ as shown in the following result[4]:

<u>Theorem 8.1</u>. Let $\alpha \in (0,1)$, $c \in (1,\infty)$, $T \in (0,\infty)$. Let $\beta : \bar{B} \to [0,1]$ satisfy $\beta(x) \leq \max\{\alpha, 1 - V(x)/\bar{V}\} \leq 1$ and let $\gamma : \bar{B} \to [1,c]$. If there exists a z such that:

(i) $x(t,z,x_0) \in \gamma(x_0)B(x_0)$ for all $x_0 \in \bar{B}$, all $t \in [0,T]$

(ii) $x(T,z,x_0) \in \beta(x_0)B(x_0)$ for all $x_0 \in \bar{B}$,

then:

(a) $x(t,z,x_0) \in c\bar{B}$ for all $x_0 \in \bar{B}$, all $t \in [0,\infty)$,

(b) $x(t,z,x_0) \to 0$ as $t \to \infty$ for all $x_0 \in \bar{B}$.

Comment: The hypotheses of Theorem 8.1 are satisfied if the function β is replaced by a constant in the interval $(0,1)$ and the function γ is replaced by the constant c. The result follows from the fact that

$$x_i \in \gamma_{i-1}\gamma_{i-2} \cdots \gamma_0 B(x_0)$$

where $x_i \triangleq x(iT,z,x_0)$ and $\gamma_i \triangleq \gamma(x_i)$, $i = 0,1,2,\ldots$

It follows from Theorem 8.1 that stability is ensured if two infinite dimensional inequalities (hypotheses (i) and (ii)) are satisfied.

Performance has to be assessed in the time domain. Suppose that the set of inputs of interest is $\{r_\alpha | \alpha \in A\}$ (e.g. $r_\alpha(t) = \alpha_0 + \alpha_1 t$) and that the instantaneous tracking error due to an input r_α is $e(t,z,x_0,\alpha)$. Then a possible performance constraint is $|e(t,z,x_0,\alpha)| \leq a(t,x_0,\alpha)$ for all $t \in [0,t_1]$, all $x_0 \in X$, all $\alpha \in A$. This is clearly an infinite dimensional inequality.

8.4 Semi-infinite Programming

It is thus apparent that the task of choosing suitable parameters for a controller (as contrasted with the task of choosing a controller structure) may in many cases be expressed as:

P1 Determine a z such that:

$$g^j(z) \leq 0, \quad j = 1,\ldots,m \tag{8.5}$$

$$\phi^j(z,\alpha) \leq 0 \text{ for all } \alpha \in A,\ j = 1,\ldots,p \tag{8.6}$$

or

P2 Determine a z which minimizes $f(z)$ subject to constraints (8.5) and (8.6).

In P2 $g^j(z) \leq 0$ is a conventional constraint and f represents some criterion (e.g. a measure of tracking error) which it is desired to minimize. If tuning is permitted then P1 is replaced by:

P3 Determine a z such that (8.5) is satisfied and such that for all $\alpha \in A$ there exists a $q \in Q$ such that:

$$\phi^j(z + r(q), \alpha) \leq 0, \quad j = 1,\ldots,p$$

Problem P2 is similarly modified if tuning is permitted. Several classes of algorithms have recently been developed for these problems. We give a brief exposition of these algorithms for the case when there are no conventional constraints and one infinite dimensional case.

8.4.1 Feasible-directions algorithms[5,6]

The constraint that we consider is

$$\phi(z,\alpha) \leq 0 \quad \text{for all} \quad \alpha \in A \tag{8.7}$$

The feasible directions algorithms are restricted to the case when A is an interval $[\alpha_0, \alpha_c]$ of the real line (e.g. time or frequency intervals). For all z let $\psi : R^r \rightarrow R$ and $\psi(\cdot)_+$ be defined by:

$$\psi(z) \triangleq \max\{\phi(z,\alpha) \mid \alpha \in A\}$$

and

$$\psi(z)_+ \triangleq \max\{0, \psi(z)\}$$

Let the feasible set (the set of z satisfying (8.7)) be denoted by F. Then:

$$F = \{z \in R^n \mid \psi(z) \leq 0\}$$
$$= \{z \in R^n \mid \psi(z)_+ = 0\}$$

All semi-infinite mathematical programmes solve an (infinite) sequence of finite dimensional subproblems. The feasible directions algorithms accomplish this as follows. For all $\varepsilon > 0$, all z let $A_\varepsilon(z)$ and $\tilde{A}_\varepsilon(z)$ be defined by:

$$A_\varepsilon(z) \triangleq \{\alpha \in A \mid \phi(z,\alpha) \geq \psi(z)_+ - \varepsilon\}$$

and

$$\tilde{A}_\varepsilon(z) \triangleq \{\alpha \in A_\varepsilon(z) \,|\, \alpha \text{ is a local minimizer of } \phi(z,\cdot) \text{ on } A\}$$

The reasonable assumptions that ϕ is continuously differentiable and that the set $\tilde{A}_\varepsilon(z)$ is finite for all z and all $\varepsilon \geq 0$ are made. For P1 a search direction is chosen which is a descent direction for $\phi(z,\alpha)$ at every $\alpha \in A_\varepsilon(z)$. Specifically the search direction $s_\varepsilon(z)$ solves:

$$\theta_\varepsilon(z) = \min_s \max_g \{\langle g,s\rangle \,|\, s \in S,\ g \in \partial_\varepsilon\psi(z)\}$$

or

$$\theta_\varepsilon(z) = \min\{||s|| \,|\, s \in \text{convex hull of } \partial_\varepsilon\psi(z)\}$$

where

$$S \triangleq \{s \,|\, ||s|| \leq 1\}$$

and

$$\partial_\varepsilon\psi(z) \triangleq \{\nabla_z\phi(z,\alpha) \,|\, \alpha \in \tilde{A}_\varepsilon(z)\}$$

and ε is chosen so that:

$$\theta_\varepsilon(z) \leq -\varepsilon$$

The step length is chosen, using an Armijo procedure, to reduce ψ. For P2 the search direction solves:

$$\theta_\varepsilon(z) = \min_{s\in S} \max\{\langle\nabla f(z), s\rangle - \gamma\psi(z)_+ ;\ \langle g,s\rangle,\ g \in \partial_\varepsilon\psi(z)\}$$

and the step length is chosen to reduce ψ if z is not feasible and to reduce f subject to the constraint if z is feasible. The positive constant γ ensures that the search direction is a descent direction for ψ when z is infeasible $(\psi(z)_+ > 0)$ and a feasible direction otherwise. In the implementable versions of these algorithms $\psi(z)$ and $\tilde{A}_\varepsilon(z)$ (the set of ε-most-active local maximizers) are only approximately evaluated, the accuracy being increased automatically to ensure convergence.

A quadratically convergent version of the algorithm is described elsewhere[7]; it uses the extension of Newton's methods for solving inequalities.

8.4.2 Outer approximations algorithms[8]

In these algorithms the semi-infinite program P1 and P2 are replaced by an infinite sequence of conventional programs in which the infinite dimensional set A (which is now not necessarily an interval) is replaced by an infinite sequence $\{A_i,\ i = 0,1,2,\ldots\}$ of suitably chosen finite subsets. The corresponding feasible sets $\{F_{A_i}\}$ are defined for all i by:

$$F_{A_i} \triangleq \{z \mid \phi(z,\alpha) \leq 0 \text{ for all } \alpha \in A_i\}$$

Since F is a subset of F_{A_i}, F_{A_i} is called an outer approximation. At iteration i, $F_{A_{i+1}}$ is formed by adding to F_{A_i} the α which (approximately) solves the global maximization problem $\max\{\phi(z,\alpha) \mid \alpha \in A\}$ and discards other elements of F_{A_i} judged to be unnecessary[7]. To solve P1 the outer approximation algorithm determines a sequence $\{z_i\}$ such that z_i approximately solves the problem of determining a z in F_{A_i}; to solve P2 z_i approximately solves the problem of minimizing f(z) subject to the constraint that z lies in F_{A_i}.

8.4.3 Non-differentiability[9,10,11]

In some problems (e.g. those with singular value constraints) ϕ is not differentiable with respect to z (for given α); algorithms have been developed[9,10] for the case when ϕ (and therefore ψ) is Lipschitz continuous. These algorithms are similar to those described above with $\partial_\varepsilon\psi$ now defined to be the smeared generalised derivative:

$$\partial_\varepsilon\psi(z) \triangleq \cup \{\partial\psi(z') \mid \|z'-z\| \leq \varepsilon\}$$

where the generalized derivative $\partial\psi(z)$ is defined to the convex hull of the limit of all sequences $\{\partial\psi(z_i)/\partial z\}$ where

$z_i \to z$ is such that $\partial\psi(z_i)/\partial z$ exists (ψ is differentiable almost everywhere). Since $\partial_\varepsilon\psi(z)$ is a set and not computable, it has to be approximated by the convex hull of a finite number of its elements. An adequate approximation can be achieved with a finite number of operations[9,10,11]. To solve P1 the search direction is $-h_\varepsilon(z)$ where $h_\varepsilon(z)$ is that point in $\partial_\varepsilon\psi(z)$ (or its finite approximation) closest to the origin. Corresponding extensions to solve P2 together with an application to satisfying constraints on singular values of transfer functions over an interval of frequencies is described in Reference 10. Further non-differentiable optimization algorithms are described in Reference 11.

8.4.4 Cut-map algorithms[12,13]

The cut map algorithms are outer approximation algorithms in which the outer approximations are particularly simple. Suppose there exists a (lower semi) continuous function $\delta : F^c \to R$ such that $\delta(z) > 0$ and $B(z, \delta(z)) \triangleq \{z' \mid ||z'-z|| < \delta(z)\}$ does not intersect F for all z in F^c. Then $B(z, \delta(z))^c$ is an outer approximation to F and the map $z \to B(z, \delta(z))^c$ is a cut map. The basic cut map algorithm[12] for solving P1 computes at iteration i any z_i not lying in $B(z_j, \delta(z_j))$, $j = 0,1,\ldots,i-1$; in fact this set may be replaced by $B(z_j, \delta(z_j))$, $j \in J(i) \subset \{0,1,\ldots i-1\}$ where $J(i)$ may be a small subset of $\{0,1,\ldots,i-1\}$.[12] The algorithm has been applied to the tolerancing problem[12] (determine a z such that $\phi(z+\alpha) \leq 0$ for all $\alpha \in A$) and to the tuning problem[13] (determine a z such that for all $\alpha \in A$ there exists a $q \in Q$ such that $\phi(z+\alpha+r(q)) \leq 0$) in circuit design and may well be applicable to the more general problems arising in control system design.

8.5 Conclusion

Many control design problems decompose into two phases: choosing a control structure and choosing an optimal set of parameters given the control structure. A design process may pass repeatedly through these two phases. The parameters are chosen to satisfy a set of inequalities specifying design objectives or to minimize a criterion subject to those inequalities. The inequalities are often infinite dimensional giving rise to semi infinite programming problems. Typical design objectives giving rise to such inequalities and methods for solving the resultant mathematical programs have been briefly presented. To apply the algorithm it is, of course, necessary to have an interactive program with the ability to evaluate $\phi(z, \alpha)$ and $\nabla_z\phi(z, \alpha)$. How this may be done for constraints on time and frequency responses is indicated elsewhere[14]. The approaches outlined here may be the only practical path towards satisfying hard constraints on time and frequency response (such as singular values).

References

1. MUNRO, N. (ed.) : 'Modern approaches to control system design', Peter Peregrinus Ltd., 1979.
2. 'Special Issue on Linear Multivariable Control Systems', IEEE Trans. AC-26, Feb. 1981.
3. ZAKIAN, V. and U. AL-NAIB : 'Design of dynamical and control systems by a method of inequalities', Proc. IEE, 120, (1973), pp. 1421-1427.
4. MAYNE, D.Q. and E. POLAK : 'Design of nonlinear feedback controllers', IEEE Trans., AC-26, (1981), pp.730-732. Also Proc. 19th Conf. on Decision and Control, Albuquerque, New Mexico, pp. 1100-1103 (1980).
5. MAYNE, D.Q. and E. POLAK : 'An algorithm for optimization problems with functional inequality constraints', IEEE Trans. Auto. Control, AC-21, (1976), p. 184.
6. MAYNE, D.Q., E. POLAK and R. TRAHAN : 'Combined phase I, phase II methods of feasible directions', Mathematical Programming, Vol. 17, (1979), pp. 61-73.
7. MAYNE, D.Q. and E. POLAK : 'A quadratically convergent algorithm for solving infinite dimension inequalities', submitted to Applied Mathematics and Optimization.
8. MAYNE, D.Q., E. POLAK and R. TRAHAN : 'An outer approximations algorithm for computer aided design problems', J. of Optimization Theory and Applications, 28, (1979), pp. 231-252.
9. MAYNE, D.Q. and E. POLAK : 'On the solution of singular value inequalities over a continuum of frequencies', IEEE Trans., AC-26, (1981), pp. 690-694. Also Proc. 19th

Conf. on Decision and Control, Albuquerque, New Mexico, (1980), pp. 23-28.

10. MAYNE, D.Q. and E. POLAK : 'Algorithms for the design of control systems subject to singular value inequalities', submitted to Mathematical Programming Studies.

11. MAYNE, D.Q., E. POLAK and Y. WARDI : 'Extension of non-differentiable optimization algorithms for constrained optimization', to be submitted to SIAM J. of Control and Optimization.

12. MAYNE, D.Q., A. VOREADIS : 'A cut map algorithm for design problems with parameter tolerance', accepted by IEEE Trans. on Circuit Theory.

13. MAYNE, D.Q. and A. VOREADIS : 'A cut map algorithm for design problems with parameter tolerances and tuning' submitted to IEEE Trans. on Circuit Theory.

14. MAYNE, D.Q., R.G. BECKER, A.J. HEUNIS : 'Computer aided design of control systems via optimization', Proc. IEE, Vol. 126, (1979), pp. 573-578.

Chapter 9

Optimisation in multivariable design

Dr. G. F. BRYANT

Synopsis

Three new basic algorithms to achieve diagonal dominance will be described in this paper: (i) to choose optimum permutation of input or output variables of the plant; (ii) the optimal choice of scaling factors, and (iii) to find a real precompensator to optimise dominance measure of the compensated plant.

9.1 Introduction

The design of multivariable systems via frequency domain based procedures has been intensively studied in the last decade and an array of significant theoretical results and design procedures have emerged[1]. For non trivial problems the designer invariably requires the aid of a computer and in practice CAD facilities in which the designer can 'interact' with the design steps have played a significant role. As yet 'black and white' displays have proved adequate, but it seems likely that 'colour' displays will soon prove of value and that formal optimisation procedures will play an increasingly important role as a design aid.

One of our most widely used and studied design procedures employs the concept of dominance of the plant transfer function[1,5] and we will mainly concentrate on the use of formal optimisation procedures in this area. Three important techniques can be employed to obtain a successful design. These are:

9.1.1 The allocation problem

Swap the inputs or outputs of the plant to obtain a better 'pairing' for subsequent precompensator design and design of single input-single output closed loops. The pairing that may emerge from physical considerations or during model formulation is certainly not always the best from a control point of view and we will consider which pairing is 'best' in terms of dominance. The technique is of general value in design, and not limited to those based on dominance. We will show that modest sized problems can be solved 'by eye' using simple calculations, and larger problems require graph-theoretic methods.

9.1.2 The scaling problem

Scaling of inputs and outputs changes dominance, and dominance based design approaches might be criticised by not being 'scale invariant'. As can be shown[4], 'optimal' scaling yields 'invariant' dominance, which can be calculated using linear programming based methods. We will describe algorithms for calculating optimal scaling.

9.1.3 Precompensation design problem

The calculation of 'real' compensators which are optimum over the bandwidth of the plant, has proved one of the most difficult obstacles in the application of dominance[5] and a large number of heuristic solutions are described in the literature. As the methods we describe usually yield 'global' optima' they usually serve in practice to settle the question of whether dominance is attainable for the given class of compensator.

9.2 Problem Formulation

A matrix $Q \in C^{nxn}$ is said to be column dominant if

$$|q_{ii}| \geq \delta_i^c$$

where

$$\delta_i^c = \sum_{\substack{j=1 \\ j\neq i}}^{n} |q_{ji}| \qquad i = 1,2,\ldots,n$$

and row dominant if

$$|q_{ii}| \geq \delta_i^r$$

where

$$\delta_i^r = \sum_{\substack{j=1 \\ j\neq i}}^{n} |q_{ij}| \qquad i = 1,2,\ldots,n$$

The matrix $Q \in C^{nxn}$ is said to be dominant if either of the above inequalities hold, and strictly dominant if either hold strictly. In multivariable design applications[1] it is necessary to consider transfer function matrix $Q(s) \in C^{nxn}$, when each of the elements are usually rational functions of a complex parameter $s \in C^1$. If D is a suitable Nyquist contour in the extended complex s-plane, we say Q is (strictly) dominant on D if Q(s) is (strictly) dominant for all s ϵ D .

9.2.1 Decomposition

The simplest and most frequently used precompensator design problem can be stated as follows. Given a square transfer function matrix $G(s) \in C^{nxn}$ find a real precompensator $K \in R^{nxn}$ such that at frequency $s = j\omega$ say,

$$Q(j\omega) = G(j\omega)K$$

is dominant. Selecting the column dominance criteria we require

$$|q_{ii}| \geq \delta_i^c \qquad i = 1,2,\ldots,n \tag{9.1}$$

Obviously

$$q_{ij} = \sum_{p=1}^{n} g_{ip}(j\omega)k_{pj}$$

or in short form $q_{ij} = G_{i*}(j\omega)K_{*j}$, where $G_{i*}(G_{*i})$ is the ith row (column) vector of G. It is evident that the ith column of Q is determined by the ith column of K and hence we may without loss of generality separately consider the choice of each of the n columns of K when seeking to satisfy inequality (9.1) . This decomposition of the problem into n separate problems is of great practical value.

9.2.2 Choice of dominance measure

We first note that both $|q_{ii}|$ and δ_i^c are positive first order homogeneous functions of K in the sense that for any $\lambda \neq 0$

$$|q_{ii}(\lambda K)| = |\lambda| q_{ii}(K) \tag{9.2}$$

and

$$\delta_i^c(\lambda K) = |\lambda| \delta_i^c(K) \qquad i = 1,2,\ldots,n$$

Hence a suitable and normalised dominance measure widely adopted in the literature[3] is

$$J_i^c = \frac{\delta_i^c}{|q_{ii}|} \tag{9.3}$$

where for dominance of column i we require $J_i^c < 1$, $(i = 1, 2,\dots,n)$.

From equation (9.2) it is clear that J_i^c is independent of λ and that J_i^c is constant on any ray in R^n and indeterminate at the origin. However, in any design the magnitude of the scaling factor is determined from engineering considerations, and the sign of λ from stability considerations, and hence these indeterminacies are acceptable. To avoid computational difficulties associated with constancy of $J_i^c(\lambda K)$ for any $\lambda \neq 0$ and the singular behaviour for $\lambda = 0$; it is usually convenient to introduce a constraint of the form $||K|| = 1$ say, where $||\cdot||$ is any convenient norm; if we also wish to avoid ambiguity in the sign of λ then a stronger constraint such as $k_{ii} = 1$ is convenient.

9.2.3 General problem

A great interest has been placed on achieving dominance over a set of frequencies $\Omega \triangleq \{\omega_i;\ i = 1,2,\dots,m\}$ usually spread through the bandwidth of the system. We need to decide a suitable form of cost function. Now design difficulties are usually reflected by the 'Worst Case' on Chebyshev measure, i.e.

$$\max_i J^c(\omega_i)$$

there generally being little value in obtaining 'tight' dominance at one frequency $(J_i^c << 1)$ if we have $J_i^c \simeq 1$ at some other frequencies. Hence, for multiple frequency case we consider the decomposed problems.

$$P_o \ : \ \min_k \max_r J^c(\omega_r); \qquad r = 1,2,\dots,m$$

$$J^c(\omega_r) = \delta_i^c(\omega_r)/|q_i^i(\omega_r)|$$

$$\delta_i^c(\omega_r) = \sum_{\substack{j=1 \\ j\neq i}}^{n} |q_j^i(\omega_r)|$$

$$q^i(\omega_r) = G(\omega_r)K$$

where $q^i \in C^{nx1}$, $K \in R^{nx1}$ and we have, where there is no ambiguity dropped the suffix i. Although all the algorithms described have been successfully extended to multiple frequency case[4], we will, for reasons of space, only state the single frequency algorithms.

9.3 Allocation Problem

Consider a single frequency problem where we seek to permutate the row (column) of a matrix to obtain (in some sense) best overall column (row) dominance of the resultant matrix. Essentially we are deciding how best to pair inputs and outputs. Although this problem is trivial for the 2x2 problems which pervade the literature, it rapidly becomes more difficult as the dimensions increase (for n = 5 there are 120 possible permutations and for n = 10 more than 3.6 million). Furthermore, studies indicate that correct pairing can significantly improve the design problem.

As each row permutation affects two column dominance (similarly for column permutation on row dominance), we need first to introduce a method of comparing the column dominance for any permutation. For any $Z \in C^{nxn}$ consider a positive matrix D^c where

$$d^c_{ij} \triangleq \frac{\sum_{\substack{p=1 \\ p\neq i}}^{n} |z_{pi}|}{|z_{ij}|}$$

It is easily shown that this represents the appropriate dominance measure (analogous to J^c_i of equation 9.3 when rows i and j are interchanged, and hence element j of column i is moved into the diagonal position i. Following the Chebyshev norm approach we require that the optimally permuted matrix displays the smallest possible maximum dominance measure. This criteria in principle only fixes one diagonal element and the dominance measure of one column. Extending Chebyshev criteria to the choice of a second diagonal element should then minimise the maximum dominance measure of the remaining columns; proceeding in this way we see the optimum permutation will select diagonal element so as to satisfy a sequence of n minmax criteria. In short, we are required to solve a sequence of problems of the form

$$P_1 : \min_p \max_i \{J_i^c[P,Q]\}$$

An equivalent problem to P_1 is that of choosing a set of n elements of D^c which form a 'perfect matching' of D^c and additionally minimise the maximum element in the matching. A set of n elements forms a perfect matching of $D^c \in \mathcal{R}^{nxn}$ if there is exactly one element of the set in every row and column of D^c. A simple heuristic for finding such a set (which usually suffices for 'by eye' solution of practical problems) is as follows:

Algorithm 1 (to find P)

Data: $D^c \in \mathcal{R}^{nxn}$

Step 0: Set $P^1 = D^c$; $\ell = 1$

Step 1: 'Colour' the smallest 'uncoloured' element of P

Step 2: As each additional element is coloured, check to see if the coloured set contains a perfect matching. If not, return to Step 1.

Step 3: Colour the element in D^c corresponding to the last element coloured in P^ℓ. If $\ell = n$, stop. Otherwise form a new matrix $P^{\ell-1} \in \mathcal{R}^{(\ell-1)x(\ell-1)}$ by removing the row and column of P^ℓ containing the last coloured element. Set $\ell = \ell+1$. Return to Step 2.

The sequence of coloured element in D^c is the optimal solution. To see this, consider the first time Step 3 is executed. The last element coloured is the smallest element that can be chosen in its row and column, whilst at the same time ensuring the other rows and columns can be 'covered' by the set already coloured and comprising elements of D^c smaller than the last element coloured. The same reasoning applies for the successively reduced matrices P^ℓ each time we enter Step 3; hence we end up with a perfect cover whose elements also satisfy the sequence of minmax properties required for optimality. We note that the solution is unique if the elements of D^c are distinct.

The only step in the algorithm where difficulty can arise is in checking to see if the coloured elements contain a matching (Step 2), and usually this is possible 'by eye' for multivariable design problems ($n \leq 3$?). For larger problems there is advantage in finding the optimum match automatically, especially since it may be required several times in a design (as shown by example later). The graph-theoretic equivalent of the problem is a minimax 'bottleneck' matching problem for which powerful algorithms exist. For instance modern bipartite graph theory matching algorithms or simple modification of the classical 'Hungarian' algorithm may be used[6].

9.4 Scaling Problem

Given transfer function $G \in \mathcal{C}^{n\times n}(s)$, for a linear square system with zero initial condition

$$y(s) \quad = \quad Gu(s)$$

where $y \in \mathcal{C}^{n\times 1}$ is the output and $u \in \mathcal{C}^{n\times 1}$ is the input and s is the Laplace Transform. If the outputs y_i are each scaled by factor s_i respectively $(i = 1,2,\ldots,n)$, then the transfer function is simply

$$Z \quad = \quad SG$$

where $Z \in \mathcal{C}^{n\times n}(s)$, and $S = \mathrm{diag}\{s_i\}$, $s_i > 0$. The fact that dominance is not 'scale invariant' has at least represented a philosophical criticism of dominance based approaches.

Again the problem may be easy for $n = 2$, but rapidly becomes difficult for $n > 2$. Since the worst of the column dominances represents a measure of design difficulty, it is again natural to adopt a Chebyshev norm optimality criteria and seek

$$\min_{[s]} \max_{i} \{J_i^c(Z)\}$$

For convenience, introducing the matrix M

$$m_{ij} = |z_{ij}|/|z_{ii}| \qquad i \neq j$$
$$= 0 \qquad \text{otherwise}$$

and define $\eta = \max_i \{J_i^c(Z)\}$

(and taking into account the requirement $s_i > 0$, $i = 1,2,\ldots,n$), we seek a solution of the problem

$$P_2 : \min_s \{\eta\}$$

subject to:
$$\sum_{\substack{j=1 \\ j\neq i}}^{s} m_{ji} s_j \leq \eta \, s_i \, ; \qquad i = 1,2,\ldots,n$$
$$s_i > 0 \, , \qquad i = 1,2,\ldots,n$$
$$\sum_{i=1}^{n} s_i = 1$$

where the last equality is introduced to remove indeterminacy with respect to scaling in S.

A simple procedure for solving the optimization problem is as follows:

<u>Algorithm 2</u>

Data: $\varepsilon > 0$

Step 0: Set $\ell = 1$; $\eta_1 = \max_i \sum_{\substack{j=1 \\ j\neq i}}^{n} m_{ji}$

Step 1: Solve $\min_s \{z^\ell\}$

Subject to: $(M^T - \eta_\ell I)S - z^\ell I \leq \emptyset$

$$\sum_{j=1}^{n} s_j = 1$$
$$s_j \geq 0, \qquad j = 1,2,\ldots,n$$
$$z^\ell \geq 0$$

Step 2: $d = \eta_1/2^\ell$; if $d < \varepsilon$ then stop.

Step 3: If $z^{\ell} = 0$ then $\eta_{\ell+1} = \eta_{\ell} - d$

else $\eta_{\ell+1} = \eta_{\ell} + d$

Step 4: $\ell = \ell+1$. Go to Step 1.

The scalar ε determines the absolute accuracy to which η is determined. Gershgorin's theorem ensures $\eta_1 > \hat{\eta}$ in Step 0 ($\hat{\eta}$, $\hat{s}$ denote optimal solution to this problem). Step 1 involves solution of a standard linear program. Step 3 represents a standard binary search procedure on the interval $[0,\eta]$ in which successive bisection search is employed to more closely locate $\hat{\eta}$.

9.5 Compensator Design

Given $Q \in C^{nxn}$, we seek a real compensator $K \in R^{nxn}$ such that

$$Z = QK$$

has a minimum normalised column dominance. We may without loss of generality, drop the suffix i and letting it equal to 1, consider n separate problems (P_3^o) of the form

$$P_3^o : \min_{[K]}\{\delta_2^c/|z_1|_2\}$$

$$z = Qk$$

$$||k|| = 1$$

with $z \in C^{nx1}$, $k \in C^{nx1}$ and if $z_j = \alpha_j + i\beta_j$

$$|z_j|_2 \triangleq \sqrt{\alpha_j^2 + \beta_j^2}, \quad j = 1,2,\ldots,n \text{ and}$$

$$\delta_2^c = \sum_{i=2}^{n} |z_i|_2 .$$

This is the 'single frequency' version, the multiple frequency case can be stated as

$$P_3^1 : \min_{k} \max_{\omega_i}\{\delta_2^c(\omega_i)/|z_1|_2\}$$

where ω_i, $i = 1,\ldots,r$ represent selection frequencies covering the system bandwidth.

The single frequency problem can be trivially re-expressed

$$P_3 : \min_{[k]} \eta$$

$$\delta_2^c - \eta||Z_1|| \leq 0 \qquad (9.4a)$$

$$Z = QK \qquad (9.4b)$$

$$||k|| = 1 \qquad (9.4c)$$

A number of procedures may be used to solve this problem. A simple approach which has proved effective is

Algorithm 3 (Conceptual)

Data: $\varepsilon > 0$

Step 0: Set $\ell = 1$, $\eta_1 = \delta_2^c(\bar{k})$, $\bar{k}_i = 1$, $i = 1,\ldots,n$

Step 1: If inequalities (9.4) are consistent, set $d = \eta_1/2^\ell$

$$\eta_{\ell+1} = \eta_\ell - d$$

otherwise set $\eta_{\ell+1} = \eta_\ell + d$

Step 2: If $d < \varepsilon$ stop, otherwise set $\ell = \ell+1$. Return to Step 1.

The initial step ensures that $\eta = \eta_1$ yields a consistent set of inequalities (9.4a). Step 1 employs a simple binary search on $[0,\eta_1]$. The consistency test in Step 1 is the only difficult step. It is well known that feasible direction algorithms may be used to solve sets of inequalities such as (9.4a), but that such algorithms may be slow[7]. A recently developed algorithm which employs both a Newton like step with quadratic convergence together with a first order step which yields finite convergence to a feasible point[2] has been utilised with considerable success. If the Newton and the first order steps fail, we assume the inequalities are inconsistent. More reliable behaviour and faster convergence was obtained when equality constraint (9.4c) was replaced by inequality constraints $1 \leq ||K|| \leq L$ say, where $L >> 1$; this is admissible in view of the optimality criteria being first order homogeneous in K. Also it is always advantageous to substitute (9.4b) into (9.4a) and hence reduce the problem to checking two simple inequalities for consistency.

It should be remarked that the multiple frequency case involves a minmax criteria and that powerful methods have also been developed for this general class of design problem[9]. These methods again utilise linear programming subroutines; for the multiple frequency case they behave like Newton methods and display quadratic convergence (whereas in the single frequency case they are analogous to steepest ascent and display only linear convergence).

Finally, mention should be made of special procedures tailored to the problem[8]. These have proved reliable but are computationally demanding.

9.6 Design Example

Here is an example of applying these new algorithms developed earlier to Direct Nyquist Method (DNA). The linear jet engine model chosen is a 5-states, 3-inputs, 3-outputs simplified model[5]. Only a simplified actuator dynamic is included, sensor dynamic is neglected and all output variables are assumed measurable, and the plant transfer matrix is normalised by the appropriate input/output nominal operating factors[5].

Achieving dominance (a multiple frequency case):

Step 1: Choose a set of frequency which will approximately cover the whole operating frequency range of the plant to be controlled.

Step 2: With A-1, to compute a permutation matrix P : $(G_1 = P_1 G)$ yields

$$P_1 = \begin{bmatrix} 0 & 0 & 1 \\ 0 & 1 & 0 \\ 1 & 0 & 0 \end{bmatrix}$$

The origin maximum dominance before is:

$$\delta^c_{max} = (51.33, \quad 527.4, \quad 0.325)$$

The improved maximum dominance after permuting is:

$$\delta^{c}_{max} = (0.0968, \quad 527.4, \quad 16.06)$$

Step 3: By A-2, output scaling algorithms with the same set of frequency: $(G_2 = S_1G_1S_1^{-1})$.

The output rescaling matrix is

$$S = \text{diag}\{0.14622, \quad 0.9134, \quad 1.9403\},$$

improved dominance measure is

$$\delta^{c}_{max} = \{7.92, \quad 0.285, \quad 7.83\}$$

Step 4: Re-permuting output again: $(G_3 = P_2G_2)$ gives

$$P_2 = \begin{bmatrix} 1 & 0 & 0 \\ 0 & 0 & 1 \\ 0 & 1 & 0 \end{bmatrix}$$

The improved dominance measure is

$$\delta^{c}_{max} = (7.92, \quad 6.85, \quad 0.194)$$

Step 5: Rescaling again with A-2 $(G_4 = S_2G_3S_2^{-1})$ gives

$$S_2 = \text{diag}\{1.72190, \quad 1.14100, \quad 0.13700\}$$

and improved dominance measure

$$\delta^{c}_{max} = \{1.85, \quad 1.85, \quad 1.85\}$$

Step 6: Compute a real pre-compensator by A-3 $(G_5 = G_4K_1)$ gives

$$K_1 = \begin{bmatrix} 1 & 5.12849E-3 & -6.2573E-3 \\ 35.131 & -1 & 1.51895 \\ 1 & -0.13686 & 1 \end{bmatrix}$$

and the improved dominance measure is

$$\delta^{c}_{max} = (0.9, \quad 0.7, \quad 0.8)$$

Up to this stage the dominance level of the jet engine is acceptable. Columns 1 and 2 are both open and closed-loop dominant, column 3 is closed-loop dominant with forward gain less than 1.5. Figure 9.3 shows the DNA plots.

The process of Step 1 to Step 5 can be summarised as follows:

$$G_D = U\ G\ L$$

where $U = S_2 P_2 S_1 P_1$ and $L = S_1^{-1} S_2^{-1} K_1$

with

$$U = \begin{bmatrix} 0 & 0 & 2.51990E-2 \\ 0.5707 & 0 & 0 \\ 0 & 0.330706 & 0 \end{bmatrix}$$

$$L = \begin{bmatrix} 39.6841 & 0.2035 & -0.24831 \\ 12.7578 & -0.3631 & 0.55160 \\ 14.5903 & -1.9969 & 14.5903 \end{bmatrix}$$

The structure of the controlled system is as shown in Fig. 9.1 . In view of steady state responses and gain margins, a combined lead-compensator and proportional-integral controller is used (K_c).

$$K_c = \text{diag}\{1, \frac{10(S+5)}{(S+50)} \cdot (1.25 + \frac{1}{0.167S}), \ (1.2 + \frac{1}{0.25S})\}$$

The forward gain factor for stable and good responses is:

$$K_a = \text{diag}\{50, \quad 0.6, \quad 1.8\}$$

The closed-loop step responses are shown in Fig. 9.2 .

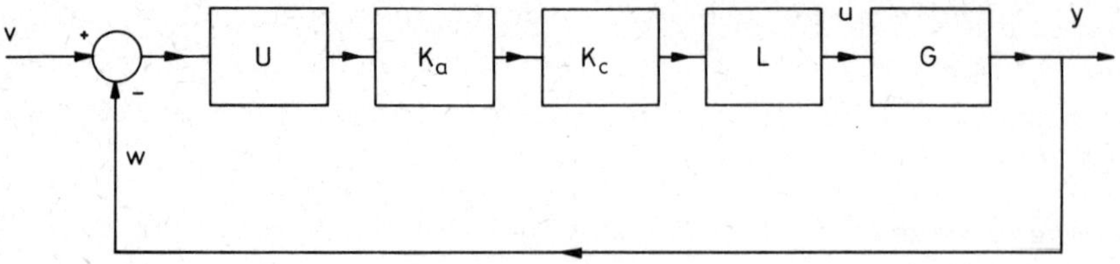

Fig. 9.1

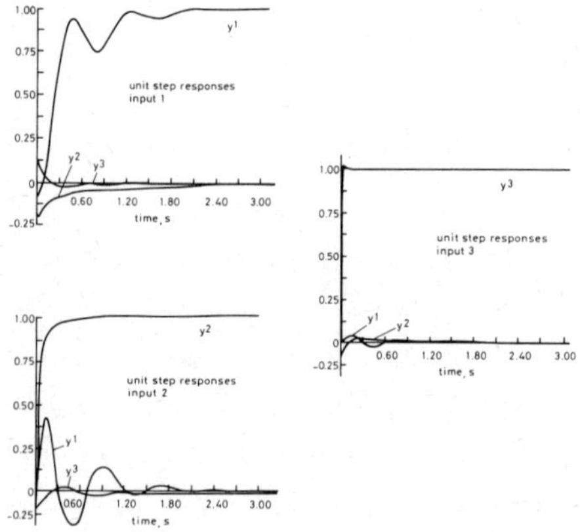

Fig. 9.2

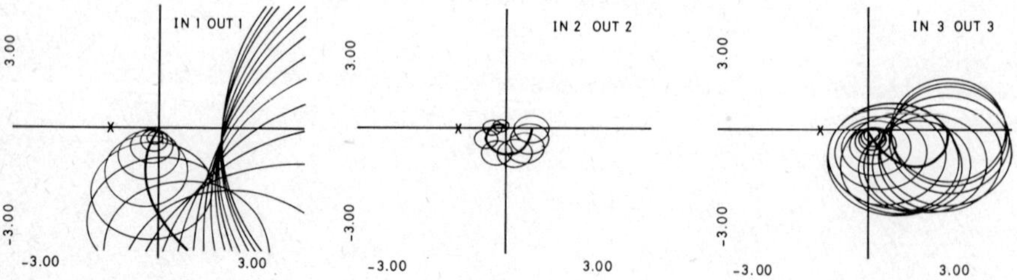

Fig. 9.3

9.7 Discussion

The algorithms described can all be extended to

(i) Multiple frequency case[4].

(ii) 'Parameterised' controllers K^c of forms such as

$$K^c = K_p + K_I/S$$

where K_p, $K_I \in R^{n\times n}$ are multivariable proportional and integral terms.

(iii) Use of inequalities on K_{ij}, $1 \leq ij \leq n$.

(iv) 'Sparse' controllers where we seek to satisfy the important practical criterion that the controllers should be of sample 'structure' and not employ complexly interconnected controllers.

(v) Solution of large sets of equations by 'tearing'. The 'tear' inputs and outputs can be 'compensated' using the methods described, and dominance is a key property for ensuring the convergence of iterated 'torn' equations.

Research is being undertaken in all these areas.

References

1. ROSENBROCK, H.H. : "Computer aided control system design", Academic Press (1974).

2. MAYNE, D.Q., E. POLAK, A.J. HEUNIS : "Solving Non-linear Inequalities in a Finite Number of Iterations", J. of Optimisation Theory and Applications, 33, No. 2, (1981).

3. KANTOR, J.C. and R.P. ANDRES : "A note on the extension of Rosenbrock's Nyquist array technique to a larger class of transfer function matrices", Int. J. Control, Vol. 30, No. 3, (1979), pp. 387-93.

4. BRYANT, G.F. : "Dominance Optimisation and Multivariable Design", Departmental Report, Dept. of Electrical Engineering, IAG Group, Imperial College (1981).

5. SAIN, M.K., J.C. PECHKOWSKI, J.L. MELSA (Eds.) : "Alternatives for linear multivariable control", National Engineering Consortium, Inc., Chicago, (1978).

6. LAWLER, G. : "Combinatorial Optimisation", Holt Reineholt (1976).

7. POLAK, E. : "Computational Methods in Optimisation", Academic Press, New York, (1971).

8. BRYANT, G.F., L.F. YEUNG : "Dominance Optimisation", IEE Control and its Applications Conference, Warwick, England, (1981).

PROBLEMS

P.9.1 Given the matrix

$$G = \begin{bmatrix} 3 & 4 & 6 & 2 & 100 \\ 400 & 30 & 200 & 1.0 & 0 \\ 0 & 1 & 4 & 60 & 10 \\ 4 & 60 & 4 & 80 & 1 \\ 0 & 1 & 1 & 30 & 1 \end{bmatrix}$$

Find the corresponding column dominance matrix D^c

$$D^c = \begin{bmatrix} 134.6 & 23 & 34 & 85.5 & .12 \\ 0.017 & 2.2 & 0.075 & 172 & \infty \\ \infty & 95 & 52.75 & 1.88 & 10.20 \\ 100.75 & 0.60 & 52.75 & 1.162 & 111.0 \\ \infty & 95 & 214 & 4.77 & 111.0 \end{bmatrix}$$

Show that the optimum permutation input → output is

$$1 \rightarrow 3, \quad 2 \rightarrow 1, \quad 3 \rightarrow 5, \quad 4 \rightarrow 2, \quad 5 \rightarrow 4$$

and confirm that the smallest five elements in D^c do not give an optimal matching. [Hint: A bigger covering set is needed.]

P.9.2 Given a system matrix $G \in R^{n\times n}$, where n = 2

$$G = \begin{bmatrix} 1 & 2 \\ 3 & 4 \end{bmatrix}$$

Find a scaling matrix $S = \mathrm{diag}\{s_i\}$, $i = 1,\ldots,n$ and $s_i \geq 0$ such that it solves the following problem:

$$P : \min_{s} \max_{j}\{\delta_j^c(S.G)\}$$

[Hint: (i) By showing that

$$J_j^c = \sum_{\substack{i=1 \\ i\neq j}}^{n} \frac{s_i}{s_j} m_{ij}, \quad \text{where} \quad m_{ij} = \frac{|g_{ij}|}{|g_{jj}|}$$

Then let $\eta = \max_j \{J_j^c(S.G)\}$

and solve the inequalities for η and S

$$\eta \geq \sum_{\substack{i=1 \\ i \neq j}}^{n} \frac{s_i}{s_j} m_{ij}, \qquad j = 1,\ldots,n$$

(ii) Hence by letting $n = 2$, and $s_1 = 1$, show that

$$s_2 = \sqrt{m_{12}/m_{21}} \qquad]$$

Chapter 10

Pole assignment

Professor N. MUNRO

Synopsis

The problem of pole assignment using state feedback or output feedback is examined. An algorithm for the design of full rank minimum degree output-feedback compensators is presented and results in a unification of previously obtained results.

10.1 Introduction

The problem of pole assignment has been extensively studied by many research workers over the last decade[1-7,9,13-16]. It is simply concerned with moving the poles (or eigenvalues) of a given time-invariant linear system to a specified set of locations in the s-plane (subject to complex pairing) by means of state or output feedback. The state-feedback approach is well established, however the output-feedback case is still being explored as a research area[18].

A large number of algorithms exist for the solution of this problem. Fortunately, these algorithms can be categorised by the characteristics of the approaches used, as shown in Figure 1. Both the state-feedback methods and the output-feedback methods result in compensator matrices which are broadly speaking either dyadic (i.e. have rank equal to one) or have full rank. Although the dyadic algorithms have considerable elegance and simplicity, the resulting closed-loop systems have poor disturbance rejection properties compared with their full-rank counterparts[19].

Here, we shall review several existing algorithms and then consider the development of an algorithm which uses the dyadic

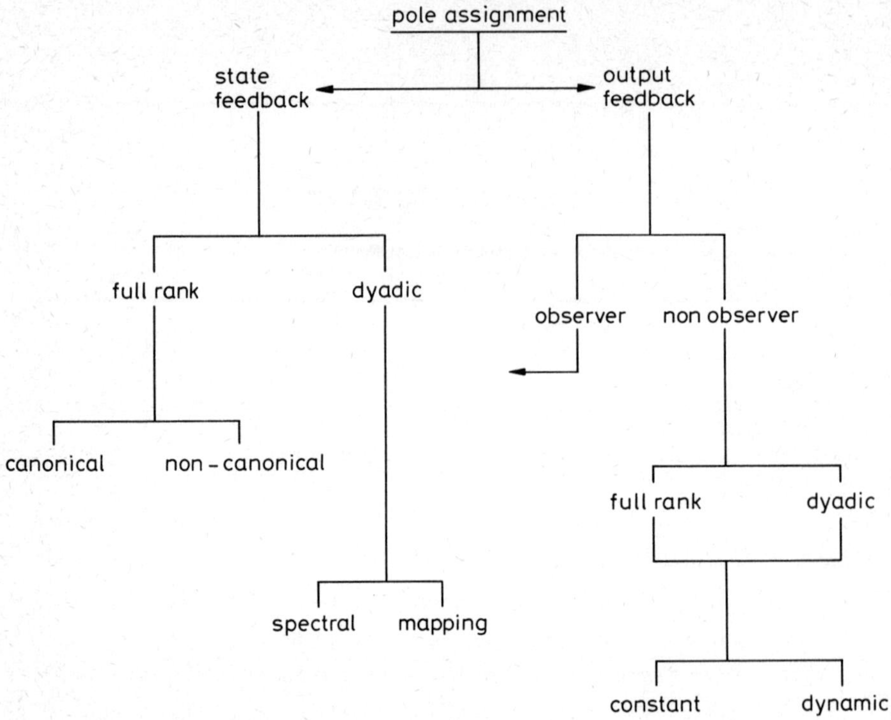

Fig 10.1 Pole Assignment Algorithms

design mechanism to generate full-rank minimum degree output-feedback compensators, which allow arbitrary pole assignment. The approach to be presented is 'consistent' with all previously established results and provides a unifying framework. The case of state-feedback is readily shown to be a special case of output-feedback.

The problem to be considered is as follows. Given a system described in state-space form as

$$\dot{\underline{x}} = A\underline{x} + B\underline{u} \tag{10.1}$$

$$\underline{y} = C\underline{x} \tag{10.2}$$

where A,B,C are real matrices with dimensions n×n, n×m, ℓ×n, respectively, or by its corresponding transfer-function matrix representation

$$G(s) = C(sI-A)^{-1}B \tag{10.3}$$

when and how can we apply feedback to this system such that the resulting closed-loop system has a desired set of arbitrary poles, or eigenvalues (subject to complex pairing). First, we consider some of the well established results, starting with the state-feedback case.

10.2 State-Feedback Algorithms

Using a feedback law of the form

$$\underline{u} = \underline{r} - K\underline{x} \tag{10.4}$$

where $\underline{r}$ is an m×1 vector of reference inputs and K is an m×n real matrix, when and how can we determine K such that the resulting closed-loop system

$$\dot{\underline{x}} = (A-BK)\underline{x} + B\underline{r} \tag{10.5}$$

has a desired set of eigenvalues $\{\gamma_i\}$; $i = 1,n$?

The question when was answered by Wonham[1] and the requirement is that the pair [A,B] is completely controllable. The question of how can be examined in several ways, as follows.

10.2.1 Dyadic designs

Under the heading of dyadic designs (see Fig. 1), the spectral approach essentially requires the calculation of the eigenvalues $\{\lambda_i\}$, and the associated eigenvectors $\{\underline{w}_i\}$ and reciprocal eigenvectors $\{\underline{v}_i^t\}$ of the matrix A. Significant contributions have been made in this area by several workers[2-5]. Here, the matrix K is considered as the outer product of two vectors; i.e.

$$K = \underline{f}\,\underline{m}^t \tag{10.6}$$

where $\underline{f}$ is m×1 and $\underline{m}^t$ is 1×n.

The problem of simultaneously determining the (m+n-1) free elements in K is nonlinear. However, by pre-assigning values to the elements of one vector, the determination of the remaining elements results in a linear problem[3]. Conventionally, the vector $\underline{f}$ is chosen such that the resulting pseudo single-input system $[A, B\underline{f}]$ is completely controllable. Then for a desired closed-loop pole set $\{\gamma_i\}$, the vector $\underline{m}^t$ is given by

$$\underline{m}^t = \sum_{i=1}^{q} \delta_i \, \underline{v}_i^t \tag{10.7}$$

where q = number of poles to be altered, the $\underline{v}_i^t$ are the reciprocal eigenvectors associated with those eigenvalues $\{\lambda_i\}$ to be altered, and the scalars δ_i are weighting factors calculated as

$$\delta_i = \frac{\prod_{j=1}^{q} (\lambda_i - \gamma_j)}{p_i \prod_{\substack{j=1 \\ j \neq i}}^{q} (\lambda_i - \lambda_j)} \tag{10.8}$$

where the scalars p_i are defined by the inner product

$$p_i = \langle \underline{v}_i , B\underline{f} \rangle \tag{10.9}$$

We note that if the λ_i or γ_i are complex then the resulting δ_i will be complex. However, $\underline{m}^t$ will be real if these values correspond to complex-conjugate pole-pairs.

When the open-loop system has any eigenvalues of multiplicity greater than 1, the above algorithm breaks down, since the denominator term in equation (10.8) becomes zero for certain δ_i. However, a modification to this algorithm to cater for this latter situation has been developed and is described in Reference [4].

In contrast to the above method, the mapping approach (see Fig. 1) due to Young and Willems[6] does not require the spectral decomposition of the open-loop system. Once the desired closed-loop system poles have been specified, the corresponding closed—loop system characteristic polynomial $\Delta_d(s)$, where

$$\Delta_d(s) = s^n + d_1 s^{n-1} + \ldots + d_n \tag{10.10}$$

can be calculated. If the open-loop system characteristic polynomial is defined as

$$\begin{aligned}\Delta_o(s) &= |sI-A| \\ &= s^n + a_1 s^{n-1} + \ldots + a_n \end{aligned} \tag{10.11}$$

then we can define a difference polynomial $\delta(s)$ as

$$\begin{aligned}\delta(s) &= \Delta_d(s) - \Delta_o(s) \\ &= s^n + (d_1-a_1)s^{n-1} + \ldots (d_n-a_n) \end{aligned} \tag{10.12}$$

Now, if the vector $\underline{f}$ is chosen such that $[A,B\underline{f}]$ is completely controllable, the required vector $\underline{m}^t$ is determined as

$$\underline{m} = [\Phi_c^t]^{-1} X^{-1} \underline{\delta} \tag{10.13}$$

where $$\Phi_c = [B\underline{f}, AB\underline{f}, \ldots, A^{n-1}B\underline{f}] \tag{10.14}$$

$$X = \begin{pmatrix} 1 & 0 & & 0 & 0 \\ a_1 & 1 & & \vdots & \vdots \\ \vdots & a_1 & & \vdots & \vdots \\ \vdots & \vdots & & \vdots & \vdots \\ \vdots & \vdots & & \vdots & \vdots \\ a_{n-2} & a_{n-1} & \cdots & 1 & \vdots \\ a_{n-1} & a_{n-2} & \cdots & a_1 & 1 \end{pmatrix} \tag{10.15}$$

and $$\underline{\delta} = [d_1-a_1, d_2-a_2, \ldots, d_n-a_n]^t \tag{10.16}$$

We note that by definition the n×n matrix X is nonsingular, and so is the controllability matrix Φ_c. Also, these need only be calculated once for a variety of closed-loop system pole specifications.

Another way of looking at this latter algorithm, which will set the scene for the results in the later sections, is from the frequency-domain point of view.

Let $$\Gamma(s) = \text{adj}(sI-A) \tag{10.17}$$

and $$\underline{b}_f = B\underline{f} \tag{10.18}$$

then, after $\underline{f}$ has been chosen to give complete controllability in the resulting single-input system, we have

$$\underline{g}(s) = \frac{1}{\Delta_o(s)}\,\Gamma(s)\,\underline{b}_f = \frac{1}{\Delta_o(s)}\begin{bmatrix} N_1(s) \\ N_2(s) \\ \cdot \\ \cdot \\ \cdot \\ N_n(s) \end{bmatrix} \tag{10.19}$$

where the $$N_i(s) = \beta_{i1}s^{n-1} + \beta_{i2}s^{n-2} + \ldots + \beta_{in} \tag{10.20}$$

and $\Delta_o(s)$ is as defined by equation (10.11).

We also have $\Delta_d(s)$ defined as

$$\begin{aligned} \Delta_d(s) &= |sI-A+BK| \\ &= |sI-A+\underline{b}_f\underline{m}^t| \\ &= |sI-A|\,|I+(sI-A)^{-1}\underline{b}_f\underline{m}^t| \end{aligned} \tag{10.21}$$

which can equally be written as

$$\Delta_d(s) = \Delta_o(s) + \underline{m}^t\,\Gamma(s)\,\underline{b}_f \tag{10.22}$$

i.e. $$\Delta_d(s) = \Delta_o(s) + \sum_{i=1}^{n} m_i\,N_i(s) \tag{10.23}$$

Using (10.12) we can consider the pole assignment problem as consisting of solving the equation

$$\sum_{i=1}^{n} m_i N_i(s) = \delta(s) \tag{10.24}$$

for the parameters of the vector $\underline{m}$.

As we shall see later, equation (10.24) is the heart of the dyadic approaches to pole assignment.

Equating coefficients of like powers in s on both sides of equation (10.24), we obtain

$$\begin{pmatrix} \beta_{11} & \cdots & \beta_{n1} \\ \vdots & & \vdots \\ \beta_{1n} & \cdots & \beta_{nn} \end{pmatrix} \begin{pmatrix} m_1 \\ \vdots \\ m_n \end{pmatrix} = \begin{pmatrix} \delta_1 \\ \vdots \\ \delta_n \end{pmatrix} \tag{10.25}$$

i.e. $$X\underline{m} = \underline{\delta} \tag{10.26}$$

where the difference vector $\underline{\delta}$ is defined by equation (10.16) and where the $n \times n$ matrix X always has full rank; i.e. the equations are consistent. Therefore, the desired vector $\underline{m}$ is given by

$$\underline{m} = X^{-1}\underline{\delta} \tag{10.27}$$

for any $\underline{\delta}$.

10.2.2 Full-rank designs

The conditions under which a full-rank state-feedback compensator K can be designed are exactly as before; namely, that the open-loop system $[A,B]$ is completely controllable. Here, again, we shall consider two distinct approaches.

In the first approach[7] we assume that the original system matrices A and B have been transformed into their so-called controllable canonical form yielding $\tilde{A}$ and $\tilde{B}$ defined as

$$\tilde{A} = TAT^{-1} \tag{10.28}$$

and $$\tilde{B} = TBQ \qquad (10.29)$$

where T is a nonsingular transformation matrix formed from specifically selected and ordered columns of the system controllability matrix $[B, AB, \ldots, A^{n-1}B]$, and Q is a nonsingular matrix formed from the nonzero rows of the matrix (TB).

Under this transformation, $\tilde{A}$ has the particular block structure

$$A = \begin{bmatrix} A_{11} & \cdots & A_{1m} \\ \vdots & & \vdots \\ A_{m1} & \cdots & A_{mm} \end{bmatrix} \qquad (10.30)$$

where the diagonal blocks A_{ii} have the companion form

$$A_{ii} = \begin{bmatrix} 0 & 1 & 0 & \cdots & 0 \\ 0 & 0 & 1 & \cdots & 0 \\ \vdots & & & & \\ 0 & 0 & 0 & \cdots & 1 \\ X & X & X & \cdots & X \end{bmatrix} \qquad (10.31)$$

with dimensions $\nu_i \times \nu_i$, where the ν_i are the controllability indices of the system; i.e.

$$\sum_{i=1}^{m} \nu_i = n$$

and where the off-diagonal blocks A_{ij}, $i \neq j$, have the form

$$A_{ij} = \begin{bmatrix} & 0 & \\ - - - & - - - & - - - \\ X & X \;\cdots & X \end{bmatrix} \qquad (10.32)$$

with dimensions $\nu_i \times \nu_j$, $i,j = 1,m$, $i \neq j$. The matrix $\tilde{B}$ has the particular block structure

$$B = \begin{bmatrix} B_1 \\ \vdots \\ B_m \end{bmatrix} \tag{10.33}$$

where

$$B_i = \begin{bmatrix} & 0 & \\ \hline 0 \; \ldots \; 0 & 1 & 0 \; \ldots \; 0 \end{bmatrix} \tag{10.34}$$

column i

has dimensions $(\nu_i \times m)$. In equations (10.31)-(10.34), the X's may be nonzero coefficients.

Given a system in this special form, the feedback matrix K required to achieve a desired closed-loop pole set $\{\gamma_i\}$ is chosen such that $(\tilde{A}-\tilde{B}\tilde{K})$ becomes block upper or lower triangular, or both, and such that the resulting diagonal blocks have nonzero coefficients which satisfy

$$\prod_{i=1}^{m} \det[sI-\tilde{A}_{ii} + (\tilde{B}\tilde{K})_{ii}] = \prod_{i=1}^{n} (s-\gamma_i) \tag{10.35}$$

The desired full-rank feedback matrix K in the original system basis is then given by

$$K = Q^{-1} \tilde{K} T \tag{10.36}$$

Consider now the situation where a full-rank feedback matrix K is to be determined as a sum of dyads; i.e. let

$$K = \sum_{i=1}^{\mu} \underline{f}^{(i)} \underline{m}^{(i)} \tag{10.37}$$

where the $\underline{f}^{(i)}$ are $m \times 1$ column vectors, the $\underline{m}^{(i)}$ are $1 \times n$ row vectors, and $\mu = \min\{m,n\} = m$. K can also be expressed as the matrix product

$$K = FM \tag{10.38}$$

where F is an m×m matrix whose columns are the $\underline{f}^{(i)}$, and M is an m×n matrix whose rows are the $\underline{m}^{(i)}$. We note that K will have full rank m, if and only if both F and M have maximum rank.

Consider the following algorithm:

(1) Suppose we choose $\underline{f}^{(1)}$ such that it selects input 1 of the given system; i.e. $\underline{f}^{(1)} = [1\ 0\ \dots\ 0]^t$. Then, since input 1 must influence $q_1 \geq 1$ poles of the system, we can determine $\underline{m}^{(1)}$ to reposition q_1 poles at desired locations.

Let the resulting closed-loop system

$$\dot{\underline{x}} = (A - B\underline{f}^{(1)}\underline{m}^{(1)})\underline{x} + B\underline{r} \tag{10.39}$$

be considered as

$$\underline{x}(s) = [\Gamma^{(1)}(s)/d^{(1)}(s)]\underline{r}(s) \tag{10.40}$$

where $$\Gamma^{(1)}(s) = \text{adj}(A-B\underline{f}^{(1)}\underline{m}^{(1)}) \tag{10.41}$$

and $d^{(1)}(s)$ is the new characteristic polynomial.

(2) Since $\Gamma^{(1)}(s)\Big|_{s=\gamma_i}$ only has rank one, we can define $\underline{f}^{(2)}$ as

$$\underline{f}^{(2)} = [f_1\ f_2\ 0\ \dots\ 0]^t \tag{10.42}$$

and can solve the equations

$$\Gamma^{(1)}(\gamma_i)\underline{f}^{(2)} = \underline{0} \tag{10.43}$$

for $f_2 \neq 0$ such that at least (q_1-1); or q_1 if $q_1 = 1$; if the poles assigned in step (1) become uncontrollable. We also note that due to the system controllability properties, this choice of $\underline{f}^{(2)}$ will influence $q_2 \geq 1$ more poles of the resulting single-input system given by

$\Gamma^{(1)}(s)\ \underline{f}^{(2)}$ and $d^{(1)}(s)$.

⋮

(m) Continuing in the above manner, it can be shown that this procedure allows us to shift all n poles of the original system to a desired set $\{\gamma_i\}$; $i = 1,n$; in m steps.

We note, that at each stage, those poles which have not been rendered uncontrollable with respect to that stage of the design procedure, and which are not being moved to desired locations will move in an arbitrary manner. However, after m stages, all n poles have been moved to the desired locations.

The matrix F generated in this manner will have full rank m, and it can also be shown that the corresponding matrix M will simultaneously have full rank n. Thus, K will have full rank; a desired objective; and will also provide the appropriate pole shifting action.

10.3 Output-Feedback Algorithms

Here, we shall adopt the frequency-domain approach from the start. Thus, given a system described by the $\ell \times m$ transfer-function matrix G(s), and using a feedback law

$$\underline{u}(s) = \underline{r}(s) - F(s)\ \underline{y}(s) \tag{10.44}$$

when and how can the resulting closed-loop system

$$H(s) = [I+G(s)F(s)]^{-1}G(s) \tag{10.45}$$

be made to have a pre-assigned set of poles $\{\gamma_i\}$; $i = 1,n$? Complete pole assignment is only possible using output-feedback when G(s) has arisen from a least-order system[8]; i.e. the corresponding state-space triple [A,B,C] is completely controllable and observable.

The following questions arise naturally at this point:

(1) What is the minimum degree of the compensator matrix $F(s)$ which will allow arbitrary pole locations in the resulting closed-loop system?
(2) What are the parameters of this $F(s)$?
(3) What is the rank of $F(s)$?

Davison has shown[9] that, using a constant feedback matrix F, $\max\{m,\ell\}$ poles can be assigned arbitrarily close to desired values. However, the remaining $n-\max\{m,\ell\}$ poles move arbitrarily. Nevertheless, this result may prove useful in situations where only a few of the open-loop system poles are to be moved to more desirable locations. It is also known that by the use of a Luenberger observer [10] of order $(n-\ell)$, or a Kalman-Bucy filter[11] of order $(n+)$, a good estimate $\hat{\underline{x}}$ of the system state-vector $\underline{x}$ can be reconstructed. This latter approach then allows any of the well known state-feedback algorithms to be applied using $\hat{\underline{x}}$. Further, if a dyadic state-feedback approach is to be used then we need only estimate $\hat{\alpha} = \underline{m}^t\underline{x}$, and this can be done under favourable conditions[12] using an observer of order $r = \nu_o-1$, where ν_o is the observability index for the system.

For those situations where a dynamic compensator $F(s)$ is required, Brasch and Pearson[13] have shown that a dynamic compensator of degree

$$r = \min\{\nu_o-1, \nu_c-1\} \tag{10.46}$$

where ν_c is the controllability index for the system, is sufficient (but not necessary) to allow arbitrary pole assignment. The design of an observer is implicit in this approach, and the resulting feedback is essentially dyadic. A similar result has been obtained by Chen and Hsu in the frequency-domain[14].

Another result, obtained by Davison and Wang[15] results in a compensator of rank two. These authors have shown that if

$$m + \ell - 1 \geq n \tag{10.47}$$

then a constant compensator F exists which allows the closed-loop system poles to be assigned arbitrarily close to a desired set $\{\gamma_i\}$.

The final result to be considered in this section is due to Munro and Novin Hirbod[16] which states that if $m + \ell - 1 \geq n$ a full-rank constant compensator F can be determined, from a minimal sequence of dyads, to assign the closed-loop system poles arbitrarily close to a predefined set $\{\gamma_i\}$. Further, if $m + \ell - 1 < n$ then a full-rank dynamic compensator $F(s)$ with degree r given by

$$r \geq \left(\frac{n-(m+\ell-1)}{\max\{m,\ell\}} \right) \qquad (10.48)$$

can be constructed from a minimal sequence of dyads, which will allow arbitrary pole assignment.

In the following, we shall consider the design of minimum degree dyadic dynamic feedback compensators, and finally the design of minimum degree full-rank dynamic feedback compensators.

10.3.1 Dyadic designs

Given a strictly-proper $\ell \times m$ transfer-function matrix $G(s)$, let the required compensator $F(s)$ be considered as the outer product of two vectors $\underline{f}$ and $\underline{m}^t(s)$; i.e.

$$F(s) = \underline{f}\,\underline{m}^t(s) \qquad (10.49)$$

As before, the constant vector $\underline{f}$ is chosen such that

$$g(s) = G(s)\,\underline{f} \qquad (10.50)$$

is completely controllable. Let $\underline{g}(s)$ be expressed as

$$\underline{g}(s) = \frac{1}{\Delta_o(s)} \begin{bmatrix} N_1(s) \\ \vdots \\ N_\ell(s) \end{bmatrix} \qquad (10.51)$$

where

$$\Delta_o(s) = s^n + a_1 s^{n-1} + \ldots + a_n \qquad (10.52)$$

is the characteristic polynomial of $G(s)$, and the

$$N_i(s) = \beta_{i1} s^{n-1} + \beta_{i2} s^{n-2} + \ldots + \beta_{in} \qquad (10.53)$$

Now, let $\underline{m}^t(s)$ have the form

$$\underline{m}^t(s) = \frac{1}{\Delta_c(s)} [M_1(s) \ldots M_\ell(s)] \qquad (10.54)$$

where

$$\Delta_c(s) = s^r + \gamma_1 s^{r-1} + \ldots + \gamma_r \qquad (10.55)$$

is the characteristic polynomial of $\underline{m}^t(s)$, and the

$$M_i(s) = \theta_{io} s^r + \theta_{i1} s^{r-1} + \ldots + \theta_{ir} \qquad (10.56)$$

Then, Chen and Hsu[14] have shown that the resulting closed-loop system characteristic polynomial $\Delta_d(s)$, where

$$\Delta_d(s) = s^{n+r} + d_1 s^{n+r-1} + \ldots + d_{n+r} \qquad (10.57)$$

is given by the relationship

$$\Delta_d(s) = \Delta_o(s)\Delta_c(s) + \sum_{i=1}^{\ell} N_i(s)M_i(s) \qquad (10.58)$$

which is the heart of the dyadic approach to pole assignment using output-feedback.

If we now define a difference polynomial $\delta_r(s)$ as

$$\delta_r(s) = \Delta_d(s) - \Delta_o(s).s^r \qquad (10.59)$$

then (10.58) can be written as

$$\{\sum_{i=1}^{\ell} N_i(s)M(s)\} + \Delta_o(s)(\Delta_c(s)-s^r) = \delta_r(s) \qquad (10.60)$$

Equating coefficients of like powers of s on both sides of equation (10.60) yields the vector-matrix equations

$$X_r \underline{p}_r = \underline{\delta}_r \tag{10.61}$$

where X is an $(n+r)\times((r+1)(\ell+1)-1)$ matrix constructed from the coefficients a_i of $\Delta_o(s)$ and the β_{ij} of the $N_i(s)$, the vector $\underline{p}_r$ contains the unknown parameters γ_j and θ_{ij} defining $F(s)$, and the difference vector $\underline{\delta}_r$ contains the coefficients δ_j of $\delta_r(s)$ as defined by equation (10.59). The elements of X_r, $\underline{p}_r$, $\underline{\delta}_r$ are shown explicitly in Appendix 1.

The pole assignment problem can now be stated as when can equation (10.61) be solved for the vector $\underline{p}_r$? A necessary and sufficient condition for equation (10.61) to have a solution for the $(\ell(r+1)+r)$ vector $\underline{p}_r$ is that

$$X_r X_r^{g_1} \underline{\delta}_r = \underline{\delta}_r \tag{10.62}$$

where $X_r^{g_1}$ is a g_1-inverse[17] of the matrix X_r defined by

$$X_r X_r^{g_1} X_r = X_r \tag{10.63}$$

If equation (10.61) is consistent, the general solution for $\underline{p}_r$ is

$$\underline{p}_r = X_r^{g_1} \underline{\delta}_r + (I - X_r^{g_1} X_r)\underline{z} \tag{10.64}$$

where $\underline{z}$ is an arbitrary $[\ell(r+1)+r]\times 1$ vector. Since $\underline{z}$ is arbitrary, it can be chosen as the zero vector yielding

$$\underline{p}_r = X_r^{g_1} \underline{\delta}_r \tag{10.65}$$

We note that if the consistency condition given by equation (10.62) is not satisfied for a particular g_1-inverse of X_r, then it will not be satisfied for any other g_1-inverse of X_r.

It has already been mentioned that if r is chosen as

$$r_m = \min\{\nu_o-1,\nu_c-1\} \tag{10.66}$$

hen arbitrary pole assignment can be achieved. This result

can now be simply explained, since for r defined by (10.66) we obtain under these special conditions

$$X_r X_r^{g_1} = I_{n+r} \tag{10.67}$$

This is due to the rank of X_r, for this particular value of r, being such that a right-inverse of X_r exists, defined by

$$X_r^{g_1} = X_r^t [X_r \; X_r^t]^{-1} \tag{10.68}$$

It is also interesting to note that for arbitrary pole assignment, r is bounded as

$$\left(\frac{n-\ell}{\ell} \right) \leq r_m \leq n-\ell \tag{10.69}$$

We can now see that for the case of state-feedback, since then $\ell = n$, equation (10.69) yields $r = 0$; i.e. F is a constant matrix.

However, equation (10.69) does not allow for known exceptions, where arbitrary pole assignment can be achieved with a compensator $F(s)$ of order $r < r_m$. To allow for these situations, we state that the necessary and sufficient conditions on r are that it is chosen such that

$$\rho\{X_r\} = \rho\{X_r, \underline{\delta}_r\} \tag{10.70}$$

where $\rho\{\cdot\}$ means rank.

10.3.2 Full-rank designs[16]

Suppose that the required compensator F, constant or dynamic, is constructed from a minimal sequence of dyads as

$$F = \sum_{i=1}^{\mu} \underline{f}^{(i)} \; \underline{m}^{(i)} \tag{10.71}$$

where $\underline{f}^{(i)}$ is an $m \times 1$ vector, $\underline{m}^{(i)}$ is a $1 \times \ell$ vector, and

$$\mu = \min\{m, \ell\} \tag{10.72}$$

Then, equation (10.71) can be equally written as

$$F = F_c F_o \qquad (10.73)$$

where F_c is an $m \times m$ matrix whose columns are the vectors $\underline{f}^{(i)}$, and where F_o is an $m \times \ell$ matrix whose rows are the vectors $\underline{m}^{(i)}$.

We can ensure that for the case where $m < \ell$ the matrix F_c has full-rank by its construction, but F will have full rank if and only if F_o and F_c both have full rank. It is tedious but straightforward[16] to show that if F_c is generated by the method shown in the following algorithm, then the corresponding matrix F_o simultaneously has full rank. The case where $m > \ell$ is obtained by duality.

Algorithm[16]

(1) Choose $\underline{f}_c^{(1)}$ such that it picks up input 1, say, of the system; i.e. let $\underline{f}_c^{(1)} = [1\ 0\ \ldots\ 0]^t$; and determine the corresponding vector $\underline{f}_o^{(1)}$ to place as many poles as possible in desired locations. Since the system is completely controllable, the pseudo-input vector $\underline{b}^{(1)} = B\underline{f}_c^{(1)}$ will influence n_1 modes of the system where $1 \leq n_1 \leq n$. Also, the system $[A, \underline{b}^{(1)}, C]$ has ℓ_1 linearly independent outputs and one input and, in fact, only $q_1 = \min\{\ell_1, n_1\}$ poles can be assigned to specific locations at this stage.

(2) For $\underline{f}_c^{(2)} = [f_1\ f_2\ 0\ \ldots\ 0]^t$, say, determine f_1 and f_2; $f_2 \neq 0$; to render as many as possible of the q_1 poles assigned in stage 1 uncontrollable, and determine $\underline{f}_o^{(2)}$ to move at least one more pole to a desired location.

Let n_2 be the number of modes influenced by $\underline{b}^{(2)} = B\underline{f}_c^{(2)}$ where $1 \leq n_2 \leq n-1$. Then, $\underline{f}_o^{(2)}$ is determined to move $q_2 = \min\{\ell_2, n_2\}$ poles to desired locations by solving the equation (10.74):

$$\Delta^{(2)}(s) = \Delta^{(1)}(s) + \underline{f}_o^{(2)} C[\mathrm{adj}(sI-A+B\underline{f}_c^{(1)}\underline{f}_o^{(1)}C)]B\underline{f}_c^{(2)}$$

where

$$\Delta^{(1)}(s) = \Delta_o(s) + \underline{f}_o^{(1)} C[\mathrm{adj}(sI-A)]B\underline{f}_c^{(1)} \qquad (10.75)$$

Note that the elements f_1 and f_2 of $\underline{f}_c^{(2)}$ are determined by solving

$$\Gamma^{(1)}(s)\underline{f}_c^{(2)} = \underline{0}$$
$$= [\mathrm{adj}(sI-A+B\underline{f}_c^{(1)}\underline{f}_o^{(1)}C)]B\underline{f}_c^{(2)} \qquad (10.76)$$

at those roots of $\Delta^{(1)}(s)$ which correspond to eigenvalues to be made uncontrollable.

⋮

(m) For $\underline{f}_c^{(m)} = [f_1\ f_2\ \ldots\ f_m]^t$ determine f_1 to f_m; $f_m \neq 0$; to hold (m-1) poles in desired locations, and determine $\underline{f}_o^{(m)}$ to place $\min\{\ell, n-m+1\}$ poles in desired locations.

Provided $m+\ell-1 \geq n$, this procedure can always be carried out. Now, if $m+\ell-1 < n$, the first m-1 stages of this procedure are carried out as before, and then the final stage is carried out using a dynamic dyadic compensator of degree r, as defined by equation (10.48) . The resulting full-rank compensator F(s) will have degree r.

10.3.3 Example

Consider a system with transfer-function matrix

$$G(s) = \frac{1}{s^4}\begin{bmatrix} s^2(s+1) & (s+1) \\ 0 & s^2(s+1) \end{bmatrix} \qquad (10.77)$$

which has eigenvalues $\{\lambda_i\} = \{0, 0, 0, 0\}$, $m = \ell = 2$, $n = 4$, and has two transmission zeros at $s = -1$.

(1) Choose $\underline{f}_c^{(1)} = [1 \quad 0]^t$, then

$$\underline{g}^{(1)}(s) = \frac{1}{\Delta_1(s)} N_1(s) = \frac{1}{s^2}\begin{bmatrix} (s+1) \\ 0 \end{bmatrix} \qquad (10.78)$$

So, the system seen through input 1 has two poles at the origin and one zero at -1. Now, since $n_1 = 2$ and $\ell_1 = 1$, we get $q_1 = 1$. Thus, we determine $\underline{f}_o^{(1)}$ to move one pole to a new location at $s = -2$, say. This is achieved by $\underline{f}_o^{(1)} = [4 \quad 0]$; i.e.

$$F_1 = \begin{bmatrix} 4 & 0 \\ 0 & 0 \end{bmatrix} \tag{10.79}$$

which results in the closed-loop system pole set $\{\lambda_i\} = \{0, 0, -2, -2\}$. So, we have not just moved one pole to $s = -2$ but have in fact moved two there. This situation is fully predictable from the resulting closed-loop system root-locus plot.

(2) Now, consider $\{\lambda_i\} = [\{\gamma_i\}$ of step 1]. Since $m+\ell-1 = 3 < n$, a dynamic compensator of degree $r = 1$, as predicted by equation (10.48), is required. Let

$$\underline{f}^{(2)}(s) = \frac{1}{s+\gamma_1}[\theta_{10}s + \theta_{11} \quad \theta_{20}s + \theta_{21}] \tag{10.80}$$

and determine $\underline{f}_c^{(2)} = [f_1 \quad f_2]^t$ such that one of the poles moved to $s = -2$ in step 1 becomes uncontrollable.

After step 1, the resulting closed-loop system is

$$H^{(1)}(s) = \frac{1}{s^2(s+2)^2}\begin{bmatrix} s^2(s+1) & (s+1) \\ 0 & (s+1)(s+2)^2 \end{bmatrix} \tag{10.81}$$

and $$\Gamma^{(1)}(-2)\underline{f}_c^{(2)} = \underline{0} \tag{10.82}$$

yields $\underline{f}_c^{(2)} = [1 \quad -4]^t$, and so

$$N^{(2)}(s) = (s+2)(s+1)\begin{bmatrix} (s-2) \\ (s+2) \end{bmatrix} \tag{10.83}$$

To place the remaining poles at say $s = -1, -3, -4, -5$, the required vector $\underline{f}_o^{(2)}(s)$ is

$$\underline{f}_o^{(2)}(s) = \frac{1}{s+1}[(20s + 37)/2 \quad (-97/8)] \tag{10.84}$$

i.e.

$$F_2(s) = \frac{1}{s+1}\begin{bmatrix} (20s+37)/2 & (-97/8) \\ -2(20s+37) & (97/2) \end{bmatrix} \qquad (10.85)$$

A root-locus plot of the resulting closed-loop system pole behaviour during this step of the compensator design shows that the effect of feedback on the origin poles is stabilizing, as was also the case with loop 1 determined in the first step.

The overall full-rank feedback compensator $F(s) = F_1+F_2(s)$ which results in the desired closed-loop system poles is

$$F(s) = \frac{1}{s+1}\begin{bmatrix} (28s+45)/2 & -97/2 \\ -2(20s+37) & 97/2 \end{bmatrix} \qquad (10.86)$$

10.4 Concluding Remarks

The full rank output-feedback algorithm, described in the previous section, provides a systematic and unified approach to the problem of pole assignment using state or output feedback. Both dyadic and full-rank compensators can be readily designed. The computational procedures required are straightforward, and advantage can be taken of 'singular-value decomposition' techniques[20] to solve the resulting sets of equations.

It is also interesting to note that the same full-rank output-feedback algorithm can be directly used to design proportional-plus-integral action controllers which allow arbitrary pole assignment in the resulting closed-loop system[21]. Here, the original system [A,B,C]-matrices are augmented to introduce the desired integral action into the open-loop system description (see Seraji[22]), and the full-rank pole assignment algorithm (of Section 10.3) is then applied to the resulting noncyclic open-loop system.

Finally, it is interesting to note that although a great deal of effort has been invested in the pole assignment problem little attention has been given to the numerical properties of these algorithms. However, recently an interest has been shown in this area by numerical analysts and some new results are beginning to emerge[23].

References

1. WONHAM, W.M. : 'On pole assignment of multi-input controllable linear systems', IEEE Trans. Aut. Control, AC-12, 1967, pp. 680-665.

2. SIMON, J.D. : 'Theory and application of modal control', Systems Research Centre Report, SRC 104-A-67-46, Case Institute of Technology, U.S.A.

3. MUNRO, N. : 'Computer aided design of multivariable control systems', PhD Thesis, 1969, UMIST, Manchester, England.

4. RETALLACK, D.G. and A.G.J. MACFARLANE : 'Pole shifting techniques for multivariable feedback systems', Proc. IEE, 1970, Vol. 117, pp. 1037-1038.

5. PORTER, B. and T.R. CROSSLEY : Modal Control : Theory and Applications (Taylor and Francis, London, 1972).

6. YOUNG, P.C. and J.C. WILLEMS : 'An approach to the linear multivariable servomechanism problem', Int. J. Control, 1972, Vol. 15, No.5, pp. 961-979.

7. MUNRO, N. and A.I. VARDULAKIS : 'Pole-shifting using output feedback', Int. J. Control, Vol. 18, 1973, pp. 1267-1273.

8. ROSENBROCK, H.H. : State Space and Multivariable Theory, (Nelson, 1970).

9. DAVISON, E.J. : 'On pole assignment in linear systems with incomplete state-feedback', IEEE Trans. Aut. Control, AC-15, 1970, pp. 348-351.

10. LUENBERGER, D.G. : 'An introduction to observers', IEEE Trans. Aut. Control, AC-16, 1971, pp. 596-602.

11. BUCY, R.S. and P.D. JOSEPH : 'Filtering for stochastic processes with applications to guidance', Wiley Interscience (New York, 1968).

12. MURDOCH, P. : 'Design of degenerate observers', IEEE Trans. Aut. Control, Vol. AC-19, 1974, pp. 441-442.

13. BRASCH, F.M. and J.B. PEARSON : 'Pole-placement using dynamic compensators', IEEE Trans. Aut. Control, AC-15, 1970, pp. 34-43.

14. CHEN, C.T. and C.H. HSU : 'Design of dynamic compensators for multivariable systems', JACC, 1971, pp.893-900.

15. DAVISON, E.J. and S.H. WANG : 'On pole assignment in linear multivariable systems using output feedback', IEEE Trans. Aut. Control, Vol. AC-20, 1975, pp. 516-518.

16. MUNRO, N. and S. NOVIN-HIRBOD : 'Pole assignment using full-rank output-feedback compensators', Int. J. System Science, Vol. 10, No. 3, 1979, pp.285-306.

17. PRINGLE, R.M. and A.A. RAYNER : Generalized inverse matrices with applications to statistics, (Griffin, London, 1971).

18. WILLEMS, J.C. and W.H. HESSELINK : 'Generic properties of the pole placement problem', 7th World Congress of IFAC, Helsinki, 1978.

19. DANIEL, R.W. : 'Rank-deficient feedback and disturbance rejection', Int. J. Control, Vol. 31, No. 3, 1980, pp. 547-555.

20. STRANG, G. : Linear algebra and its applications, (Academic Press, 1976).

21. NOVIN-HIRBOD, S. : 'Pole assignment using proportional-plus-integral output feedback control', Int. J. Control, vol. 29, No. 6, (1979), pp. 1035-1046.

22. SERAJI, H. : 'On output feedback control of linear multivariable systems', in Control System Design by Pole-Zero Assignment, ed. by F. Fallside (Academic Press, 1977).

23. FLETCHER, L.R. : 'An intermediate algorithm for pole placement by output feedback in linear multivariable control systems', Int. J. Control, Vol. 31, No. 6, 1980, pp. 1121-1136.

Appendix 1

The parameters in the vector matrix equation

$$X_r \underline{p}_r = \underline{\delta}_r$$

are shown explicitly below for the case where a dynamic feedback compensator of degree r is required:

$$
\underbrace{}_{\substack{n+r\\ \text{rows}}}
\left[\begin{array}{ccc:ccc:c:ccc}
1 & 0 \cdots & 0 & \beta_{11} & 0 \cdots & 0 & & \beta_{l1} & 0 \cdots & 0 \\
\alpha_1 & 1 & \vdots & \beta_{12} & \beta_{11} & \vdots & & \beta_{l2} & \beta_{l1} & \vdots \\
\alpha_2 & \alpha_1 & \vdots & \vdots & \beta_{12} & \vdots & & \vdots & \beta_{l2} & \vdots \\
\vdots & \alpha_2 & 0 & \vdots & & \vdots & & \vdots & \vdots & \vdots \\
\vdots & \vdots & 1 & \beta_{1n} & & 0 & --- & \beta_{ln} & \vdots & 0 \\
\alpha_n & \vdots & \alpha_1 & 0 & \beta_{1n} & \beta_{11} & & 0 & \beta_{ln} & \beta_{l1} \\
0 & \alpha_n & \alpha_2 & \vdots & 0 & \beta_{12} & & \vdots & 0 & \beta_{l2} \\
\vdots & 0 & \vdots & \vdots & \vdots & \vdots & & \vdots & \vdots & \vdots \\
\vdots & 0 & \vdots & \vdots & \vdots & \vdots & & \vdots & \vdots & \vdots \\
\underbrace{0 \quad 0 \quad \alpha_n}_{\substack{r\\ \text{columns}}} & & & \underbrace{0 \quad 0 \quad \beta_{1n}}_{\substack{r+1\\ \text{columns}}} & & & & \underbrace{0 \quad 0 \quad \beta_{ln}}_{\substack{r+1\\ \text{columns}}} & &
\end{array}\right]
\left[\begin{array}{c}
\gamma_1 \\ \vdots \\ \gamma_r \\ \hdashline \theta_{l0} \\ \vdots \\ \theta_{1r} \\ \hdashline \vdots \\ \hdashline \theta_{l0} \\ \vdots \\ \theta_{1r}
\end{array}\right]
=
\left[\begin{array}{c}
d_1 \\ d_2 \\ \vdots \\ \vdots \\ \vdots \\ \vdots \\ \vdots \\ \vdots \\ d_{n+r}
\end{array}\right]
-
\left[\begin{array}{c}
\alpha_1 \\ \alpha_2 \\ \vdots \\ \vdots \\ \alpha_n \\ 0 \\ \vdots \\ \vdots \\ 0
\end{array}\right]
$$

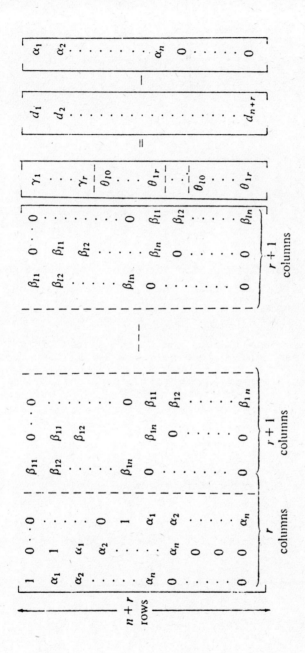

POLE ASSIGNMENT - Problems

P.1 Given a system described by the state-space equations

$$\dot{\underline{x}} = \begin{pmatrix} -2 & 1 & 2 \\ -1 & -2 & 2 \\ -2 & 0 & 2 \end{pmatrix} \underline{x} + \begin{pmatrix} 0 & 0 \\ 0 & 1 \\ 1 & 0 \end{pmatrix} \underline{u}$$

with eigenvalues $\{\lambda_i\} = \{+j, -j, -2\}$

and reciprocal eigenvectors

$$\{\underline{v}_i^t\} = \begin{pmatrix} 5 & 2-j & -6-2j \\ 5 & 2+j & -6+2j \\ 0 & -4 & 2 \end{pmatrix}$$

Use the spectral approach to determine a suitable state-feedback matrix K, such that $(A-BK)$ has eigenvalues $\{\gamma_i\} = \{-2, -2, -2\}$.

P.2 Apply Young's algorithm to the system of Problem 1 to obtain the same closed-loop pole set using state feedback.

P.3 Given a system in controllable canonical form as

$$\dot{\tilde{\underline{x}}} = \left(\begin{array}{cc|c} 0 & 1 & 0 \\ 80/9 & 26/3 & 10/3 \\ \hline 8/3 & 0 & 1/3 \end{array}\right) \underline{x} + \left(\begin{array}{cc} 0 & 0 \\ 1 & 0 \\ \hline 0 & 1 \end{array}\right) \underline{u}$$

write down a full-rank state feedback matrix $\tilde{K}$ such that $(\tilde{A}-\tilde{B}\tilde{K})$ has eigenvalues $\{\gamma_i\} = \{-4, -4, -4\}$. Given that the transformation matrices T and Q required to put the original system into controllable canonical form are

$$T = \begin{pmatrix} 0 & 0 & 1/3 \\ 1 & 4/3 & 5/3 \\ 0 & 1 & -2/3 \end{pmatrix}$$

and $$Q = \begin{pmatrix} 1 & 4/3 \\ 0 & 1 \end{pmatrix}$$

evaluate K in the original basis.

P.4 Given a system described by the transfer-function matrix

$$G(s) = \frac{1}{s^3+6s^2+7s+2}\begin{bmatrix} s^2+6s+4 \\ s+2 \end{bmatrix}$$

determine a dyadic output-feedback compensator such that the closed-loop system plant poles are $\{\gamma_i\} = \{-1, -2, -4\}$.

If the closed-loop system plant-pole locations required are now changed to $\{\gamma_i\} = \{-1, -2, -3\}$, determine a suitable dyadic output-feedback compensator.

P.5 Given

$$G(s) = \frac{1}{s^3}\begin{bmatrix} s^2+s+1 & 1 \\ s(s+1) & s \end{bmatrix}$$

determine an appropriate full-rank output feedback compensator such that the resulting closed-loop system poles are $\{\gamma_i\} = \{-1, -2, -5\}$.

Chapter 11

Nonlinear systems

Dr. P. A. COOK

Synopsis

Some methods for analysing the behaviour of nonlinear dynamical systems are discussed. Attention is concentrated on the use of describing functions for predicting limit cycles and Lyapunov's method for investigating stability. A brief treatment of absolute stability criteria and their use is also given.

11.1 Nonlinear Behaviour

The design of a control system is usually based upon a linear model of the plant to be controlled, for the good reason that the assumption of linearity makes the dynamical behaviour much easier to analyse. Moreover, the resulting design can be expected to be satisfactory at least in the neighbourhood of the operating point around which the model has been linearised. In practice, however, all systems are to some extent nonlinear and, as a result, may exhibit forms of behaviour which are not at all apparent from the study of the linearised versions. Also, in some types of control scheme (e.g. adaptive control, variable-structure systems), nonlinear features are deliberately introduced in order to improve the system performance. We begin, therefore, by considering some of the more important effects which nonlinear properties can introduce.

(i) Dependence on operating point. When a nonlinear model is linearised around a point representing steady-state operation, the result will in general depend upon the chosen operating point. Consequently, a control scheme designed for particular operating conditions may not be suitable in other circumstances. To overcome this problem, we may need to use a nonlinear controller, adapted to the nonlinearities of the plant, or, more probably, a set of different linear controllers to be used in different operating regions. Alternatively, we may try to smooth out the nonlinearities by some means, e.g. injecting a rapidly oscillating signal (dither) together with the control input.

(ii) Stability. For a linear system, local stability (i.e. in the immediate neighbourhood of an equilibrium point) is equivalent to global stability, but this is not necessarily true in the nonlinear case. Thus, even if a nonlinear system is stable for small departures from equilibrium, it may become unstable under larger perturbations, so we need to develop methods for estimating the region in which stability is guaranteed (domain of attraction). Also, instead of a unique equilibrium point, for given operating conditions, there may be many equilibria with different stability properties.

(iii) Limit cycles. A typical feature of nonlinear system behaviour is the occurrence of persistent oscillations, called limit cycles. These are normally undesirable in a control system and usually arise from the application of excessive gain, so that the local equilibrium point becomes unstable, although they can also be excited by large disturbances or changes in operating conditions. It is therefore necessary to be able to predict when they are likely to occur and to estimate their amplitudes and periods of oscillation.

(iv) Frequency response. When a linear system is given a sinusoidal input, its output is a sinusoid of the same frequency. A nonlinear system, however, will generate harmonics (at multiples of the input frequency) and possibly subharmonics (at submultiples of this frequency) or combination tones, arising from mixing of the input signal with internal oscillations of the system. It is also possible for the output amplitude to become a multivalued function of the input frequency, so that discontinuous changes (jump phenomena) may occur as the frequency is varied.

11.2 Fourier Series

If a linear element is subjected to a sinusoidal input, the output (in the steady state) is also a sinusoid of the same frequency, and this is an essential simplifying feature in the study of linear systems. For a nonlinear element, this no longer holds, and one way of describing the behaviour of such an element is to analyse the output into its component frequencies. In practice, it is often found that one particular frequency, namely the input frequency, predominates in the output to such an extent that the other components may be neglected, to a good approximation, in analysing the behaviour of the system. This is the philosophy of the *describing function* approach[1], also called the *method of harmonic balance*. It amounts to treating nonlinear systems as nearly as possible like linear systems, i.e. each frequency is treated independently, but the nonlinear behaviour shows up through the appearance of amplitude-dependent coefficients multiplying signals which pass through nonlinear elements. As a preliminary to the formulation of this method we first consider the general question of the resolution of a periodic signal into sinusoidal components (Fourier analysis).

Suppose a nonlinear element has a sinusoidal input

$$u(t) = U \sin(\omega t)$$

where U and ω are positive constants. In the case of a memoryless (i.e. single-valued) nonlinearity, or the simpler types of nonlinearity with memory, the output $y(t)$ will have the same period as the input, but will contain harmonics whose frequencies are multiples of the fundamental frequency ω. It can be represented by the *Fourier series* expansion

$$y(t) = \frac{a_0}{2} + \sum_{n=1}^{\infty} \{a_n \cos(n\omega t) + b_n \sin(n\omega t)\}$$

where the coefficients a_n, b_n are given by

$$a_n = \frac{1}{\pi} \int_0^{2\pi} y(t) \cos(n\omega t) d(\omega t)$$

$$b_n = \frac{1}{\pi} \int_0^{2\pi} y(t) \sin(n\omega t) d(\omega t)$$

as may be checked by substitution.

The constant term $a_0/2$ arises from a bias, i.e. lack of symmetry, in the nonlinearity, and will be absent if the non linearity is symmetrical; that is to say, if the output is an odd function of the input. In fact, this applies to all the *even* harmonics, i.e. those for which n = even integer. This is because, for a symmetrical nonlinearity, the output has *half-wave symmetry*, i.e.

$$y(t + \frac{\pi}{\omega}) = - y(t)$$

just like the input, and hence a_n and b_n vanish for even n, leaving only the odd harmonics (n = odd integer).

Also, if the nonlinearity is memoryless, the output has the property

$$y(\frac{\pi}{\omega} - t) = y(t)$$

again like the input, so that $a_n = 0$ for odd n and $b_n = 0$ for even n. Hence, for a symmetrical memoryless nonlinearity, the only nonzero coefficients are the b_n for odd n.

Example

An ideal relay.

With input u, this element has output

$$y = \text{sgn}\ (u)$$

i.e.

$$y = \begin{cases} 1, & u > 0 ; \\ 0, & u = 0 ; \\ -1, & u < 0 . \end{cases}$$

It is thus symmetrical and memoryless.

For the coefficients in the output, we have

$$\pi a_n = \int_0^{\pi} \cos(n\theta)d\theta - \int_{\pi}^{2\pi} \cos(n\theta)d\theta = 0$$

$$\pi b_n = \int_0^{\pi} \sin(n\theta)d\theta - \int_{\pi}^{2\pi} \sin(n\theta)d\theta$$

$$= \frac{2}{n}(1 - \cos n\pi)$$

so that

$$b_n = \begin{cases} \frac{4}{\pi n}, & n \text{ odd} ; \\ 0 \quad , & n \text{ even}. \end{cases}$$

Thus, the amplitude of the harmonics falls to zero as $n \to \infty$. This is a general feature of Fourier expansions.

11.3 The Describing Function

The idea of this approximation is to pretend that only the fundamental component in the Fourier series matters and all higher harmonics can be neglected. We shall also assume that our nonlinearities are symmetrical so that the constant bias term does not arise, and hence we approximate the signal $y(t)$ by

$$y_0(t) = a_1 \cos(\omega t) + b_1 \sin(\omega t)$$

This is actually the best linear approximation in a certain sense, since, if we define the mean-square error Δ by

$$\Delta = \int_0^{2\pi} \{y(t) - y_0(t)\}^2 \, d(\omega t)$$

and minimise this with respect to a_1 and b_1 (regarded as variable parameters), we find

$$\frac{\partial \Delta}{\partial a_1} = 2\int_0^{2\pi} \{y_0(t) - y(t)\} \cos(\omega t)d(\omega t)$$

$$= 2\{\pi a_1 - \int_0^{2\pi} y(t) \cos(\omega t)d(\omega t)\}$$

$$\frac{\partial \Delta}{\partial b_1} = 2\int_0^{2\pi} \{y_0(t) - y(t)\}\sin(\omega t)d(\omega t)$$

$$= 2\{\pi b_1 - \int_0^{2\pi} y(t) \sin(\omega t)d(\omega t)\}$$

so that Δ has a stationary point (minimum) for the values of a_1, b_1 given above.

To compare the (approximate) output with the sinusoidal input to a nonlinearity, we rewrite it as

$$y_0(t) = Y(U,\omega) \sin\{\omega t + \phi(U,\omega)\}$$

so that

$$a_1 = Y(U,\omega) \sin \phi(U,\omega)$$

$$b_1 = Y(U,\omega) \cos \phi(U,\omega)$$

where the amplitude Y and phase ϕ are functions of the input amplitude U and frequency ω. We now define the describing function

$$N(U,\omega) = \frac{Y(U,\omega)}{U} \exp\{j\phi(U,\omega)\} = \frac{b_1+ja_1}{U}$$

so that

$$y_0(t) = \mathrm{Im}\{UN(U,\omega) \exp(j\omega t)\}$$

$$u(t) = \mathrm{Im}\,\{U\exp(j\omega t)\}$$

The describing function thus represents the effective *gain* $|N|$ and *phase-shift* ϕ of the nonlinear element. For a memoryless nonlinearity, it is *real*, since $a_1 = 0$ and hence there is no phase-shift $(\sin \phi = 0)$.

11.3.1 Validity of the approximation

Using the describing function to represent any nonlinear element, we can make an approximate calculation of the behaviour of a nonlinear system, in a way which is very much simpler than solving the equations which the system actually obeys. The justification for this approximation lies in the smallness of the neglected harmonic terms, compared to the fundamental. It is therefore important that the linear elements in the system should have the property of attenuating high-frequency components more than low-frequency ones, i.e. they should act as *low-pass filters*. This is commonly the case, e.g. elements with transfer functions like

$$\frac{1}{s}, \quad \frac{1}{s^2}, \quad \frac{1}{s+1}, \quad \frac{1}{s^2+s+1}$$

would usually not spoil the describing function predictions too much. On the other hand, if the system contains an element whose gain rises with frequency, such as

$$\frac{s}{s+1}$$

or an element with a sharp resonance, like

$$\frac{1}{s^2+1}$$

then we must be wary about using the describing function, at any rate for low frequencies.

It is possible to extend the describing function method so as to take account of the presence of several components with different frequencies, whether related or not, in the same signal[2]. In control system studies, however, this is usually not necessary, at least in a semi-quantitative approximation. On the other hand, it may be important to allow for lack of symmetry in the nonlinearities (which is usually present) by incorporating bias terms in the Fourier representations. We can do this if we replace the input signal by

$$u(t) = U \sin(\omega t) + V$$

where V is a constant bias, and take the output signal representation as

$$y_0(t) = a_1 \cos(\omega t) + b_1 \sin(\omega t) + \frac{a_0}{2}$$

with the coefficients given by the formulae in Section 11.2; this again gives the best mean-square approximation possible with this parametrisation. Then, by calculating a_1, b_1 and a_0 in terms of U and V, we can obtain a generalisation of the describing function, which is now replaced by two functions (one of which may be complex to allow for a phase-shift), both depending on U, V and possibly ω.

Example

A quadratic nonlinearity,

$$y = u^2 .$$

With a biased sinusoidal input as above, the output is given by

$$\begin{aligned} y(t) &= U^2 \sin^2(\omega t) + 2UV \sin(\omega t) + V^2 \\ &= \frac{U^2}{2}\{1-\cos(2\omega t)\} + 2UV \sin(\omega t) + V^2 \end{aligned}$$

so we can immediately obtain the appropriate approximation without computing any integrals, by simply dropping the second-harmonic term. We thus have

$$y_0(t) = 2UV \sin(\omega t) + \tfrac{1}{2}(U^2 + 2V^2)$$

so that

$$a_1 = 0$$

$$b_1 = 2UV$$

$$a_0 = U^2 + 2V^2$$

and so there is no phase-shift (as expected since the nonlinearity is single-valued) but both terms in the output depend on the bias, as well as the sinusoidal amplitude, of the input. In practice, the bias would probably be essentially determined by the steady-state operating conditions and the describing function for small oscillations would be

$$N(U) = \frac{b_1}{U} = 2V$$

which is just the slope of the nonlinear function at the operating point.

11.4 Prediction of Limit Cycles

One of the principal applications of the describing function is in predicting the existence or non-existence of limit cycles in nonlinear systems. The idea is to assume that, if a limit cycle exists, it will correspond essentially to a sinusoidal oscillation of a particular frequency, so that other harmonic components can be ignored. This is exactly in the spirit of the describing function approximation and depends for its justification on the assumption that the neglected high-frequency components are effectively suppressed by the linear dynamical elements of the system. Consequently, although the predictions regarding existence and stability of limit cycles are usually fairly reliable, the calculations of amplitude and period of oscillation by this method can only be expected to have approximate validity.

Consider the behaviour of the following nonlinear feedback system, with zero reference input,

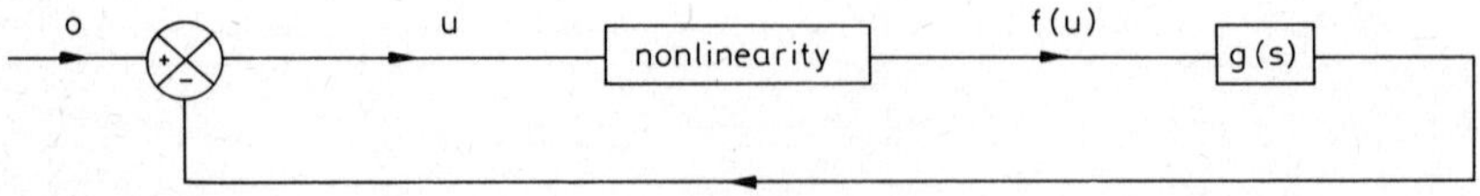

Fig. 11.1

In the describing function approximation, we assume that

$u(t)$ is a pure sinusoid,

$$u(t) = U \sin(\omega t)$$

and represent the nonlinearity by its describing function $N(U,\omega)$. The system is then described by the equation

$$U = -g(j\omega)N(U,\omega)U$$

and hence, if there is a limit cycle, i.e. a nonvanishing periodic solution, it must correspond to this equation with $U \neq 0$, so that

$$g(j\omega) = \frac{-1}{N(U,\omega)}$$

Taking the real and imaginary parts of this equation, we get two equations from which we can find U and ω, the amplitude and frequency of the limit cycle, in this approximation. If the equations have no solution, then no limit cycle is predicted.

Usually the describing function is independent of frequency and then the limit cycle condition becomes

$$g(j\omega) = \frac{-1}{N(U)}$$

In this form, it is convenient for graphical solution, by plotting $-1/N(U)$ on the Nyquist diagram of $g(s)$, and looking for intersections.

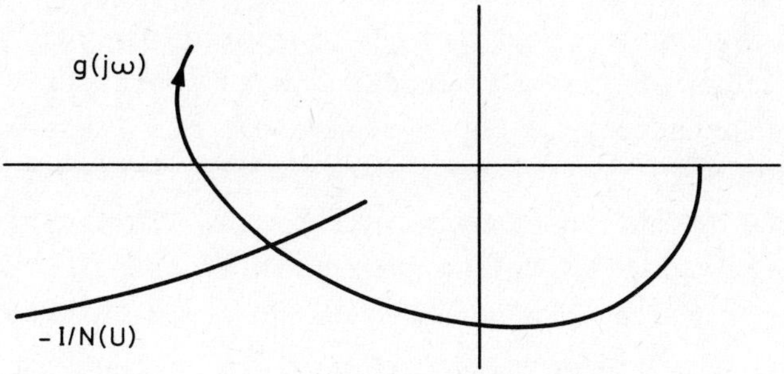

Fig. 11.2

If the graphs intersect, a limit cycle is predicted, with amplitude and frequency corresponding to the intersection point.

If there is no intersection, then some predictions about asymptotic stability, analogous to those in the linear case, may be made. Thus, if $g(s)$ is asymptotically stable, and its Nyquist plot does not encircle or intersect any part of the locus of $-1/N(U)$, we expect the feedback system to be globally asymptotically stable. All these predictions are, of course, subject to the validity of the describing function approximation, which can be relied on only under suitable conditions (in particular, the low-pass nature of the linear element).

If a limit cycle is predicted, the way in which the graphs intersect also gives us an indication of whether the limit cycle will be stable or unstable. Thus, suppose the situation is that the graph of $-1/N(U)$ in Figure 11.2 passes into the region encircled by the Nyquist plot of $g(s)$, as U increases. Assuming that $g(s)$ is stable, this indicates that an increase in U will take the system into an unstable region, thereby tending to increase U still further. Similarly, a decrease in U will make the system stable and hence cause U to diminish further. Thus, trajectories in the neighbourhood of the limit cycle will tend to diverge from it, and so the limit cycle is predicted to be *unstable*.

On the other hand, suppose the graph of $-1/N(U)$ passes out of the encircled region, as U increases. In this case, again assuming $g(s)$ to be stable, an increase in U will stabilise the system, causing U to decrease, while a decrease will destabilise it, leading to an increase in U. Hence, perturbations of the system about its limit cycling mode of operation will tend to die away as the motion returns towards the limit cycle again. Thus, the limit cycle is predicted to be *stable*. This property is important since only stable limit cycles can actually be observed in practice.

11.4.1 Frequency response of nonlinear systems

Another useful application of the describing function is in estimating the frequency response of systems containing nonlinear elements. We proceed as for linear systems, except that each nonlinear element is replaced by its describing function, in place of a transfer function. The resulting frequency response thus becomes a nonlinear function of the input amplitude and this can lead, in some cases, to the prediction of phenomena, such as jump resonance[2], which cannot occur in linear systems. Also, by extending the describing function method to take account of multiple frequencies, we can predict other essentially nonlinear phenomena, examples being subharmonic generation and the quenching of natural oscillations.

11.5 Nonlinear State-Space Equations

It is often convenient to express the equations describing a nonlinear system in state-space form, by defining a state-vector x which satisfies a first-order vector differential equation

$$\dot{x} = f(x,u,t)$$

where u is the input-vector of the system and f is a nonlinear vector function. The output-vector y of the system is then given by

$$y = h(x,u,t)$$

where h is another nonlinear vector function. If the system is time-invariant, t will not appear explicitly in the equations and, if the system is also free (independent of u) or if the input is held fixed, we can suppress the dependence on u and write

$$\dot{x} = f(x)$$

The solutions of this equation are called the state trajectories of the system.

11.6 Lyapunov's Method

This is a technique for investigating the stability properties of nonlinear state-space systems, sometimes called the second method of Lyapunov (the first method being simply linearisation around an equilibrium point). The idea is to study the behaviour of a scalar function $V(x)$, called a Lyapunov function, along the state trajectories of the system. If the state-vector x satisfies

$$\dot{x} = f(x)$$

where f is a vector function with components f_i, we have immediately, along the trajectories,

$$\dot{V}(x) = \sum_i \frac{\partial V}{\partial x_i} f_i(x)$$

which is well-defined provided that V has continuous partial derivatives with respect to all the components of x. We can then draw various conclusions from the properties of the expressions for $V(x)$ and $\dot{V}(x)$, as follows.

(i) If V is everywhere non-positive, i.e.

$$\dot{V}(x) \leq 0 \quad \text{for all } x$$

then V cannot increase along any trajectory of the system; consequently if, also, V is radially unbounded, i.e.

$$V(x) \to \infty \quad \text{as} \quad |x| \to \infty$$

where $|x|$ is the length of the vector x, then no trajectory can go off to infinity (or, equivalently, all trajectories remain bounded).

(ii) If, together with the above properties, we have

$$V(0) = 0,$$
$$V(x) > 0 \quad \text{for all} \quad x \neq 0$$

and $\dot{V}$ does not vanish identically along any trajectory of the system (apart from the degenerate trajectory $x \equiv 0$), then V must decrease to zero along every trajectory and hence all trajectories approach $x = 0$ as $t \to \infty$; the system is then said to be globally asymptotically stable.

Example

Suppose the state-space equations are

$$\dot{x}_1 = x_2$$
$$\dot{x}_2 = -x_1^3 - \gamma x_2$$

where γ is a non-negative constant, and take

$$V(x) = x_1^4 + 2x_2^2$$

so that
$$\dot{V}(x) = 4(x_1^3\dot{x}_1 + x_2\dot{x}_2)$$
$$= -4\gamma x_2^2$$

It is clear that $\dot{V} \leq 0$ everywhere and that V is radially unbounded and positive everywhere except at $x = 0$, where $V = 0$. Consequently, all trajectories remain bounded at t increases. Further, if $\gamma > 0$, $\dot{V}$ can only vanish on the line $x_2 = 0$, which is not a trajectory of the system, and so, for positive γ, the system is globally asymptotically stable.

11.6.1 Domains of attraction

If a system has an asymptotically stable equilibrium point, then, even if the stability is not global, we should be able to find a region such that all trajectories which start inside it will converge to the equilibrium point. Such a region is called a domain of attraction and one of the principal uses of Lyapunov functions lies in the determination of these domains as follows.

Suppose the equilibrium point is at $x = 0$ and that we can find a bounded region containing it, which consists of all points satisfying

$$V(x) < c$$

where c is a positive constant, such that, inside this region, V and $\dot{V}$ have all the properties required in (i) and (ii) above (except for radial unboundedness, which is now irrelevant). Then, along any trajectory starting in this region, V must decrease, which prevents the trajectory from leaving the region, and hence $V \to 0$, so that $x \to 0$, as $t \to \infty$. The region is therefore a domain of attraction. It will in general depend on the Lyapunov function used and different choices of $V(x)$ will lead to different domains being found; in this way, we may be able to construct a larger domain of attraction by taking the union of those obtained by using various Lyapunov functions.

11.6.2 Construction of Lyapunov functions

The problem of finding a suitable Lyapunov function for a given system can be tackled by various methods, none of which are entirely satisfactory. One approach is to take a function whose form is suggested by the appearance of the system's equations (e.g. an energy integral), possibly containing some arbitrary parameters, and then adjust the parameters so that V and $\dot{V}$ have the required properties.

A more systematic procedure, however, is to linearise the equations around an equilibrium point and then construct a Lyapunov function for the linearised system; this function is thus guaranteed to have the correct properties, at least in some neighbourhood of the equilibrium point. The linearised equations can be written, in vector form,

$$\dot{x} = Ax$$

where the matrix A has all its eigenvalues in the open left half-plane, since the equilibrium is assumed to be asymptotically stable. We then take the Lyapunov function as a quadratic form

$$V(x) = x^T L x$$

where suffix T denotes transposition and L is a symmetric matrix, so that

$$\dot{V}(x) = x^T(LA + A^T L)x$$

and now the correct properties for V and $\dot{V}$ can be ensured by choosing an arbitrary positive-definite matrix Q and obtaining L by solving the Lyapunov matrix equation

$$LA + A^T L = -Q$$

whose solution is the positive-definite matrix L given by

$$L = \int_0^\infty \exp(A^T t) Q \exp(At)\,dt$$

This integral expression, however, is not normally used for calculating L; usually, it is more convenient to regard the Lyapunov matrix equation as a set of linear equations for the components of L, which can be solved by standard methods of linear algebra.

When a Lyapunov function has been obtained for the linearised system, we can then apply it to the full nonlinear prolem. The expression for $\dot{V}$ will now be different but we shall still have

$$\dot{V} = -x^T Q x$$

for small $|x|$, so that there will be a region containing the equilibrium point in which $\dot{V}$ is strictly negative (except at $x = 0$). Consequently, the minimum value of V over all points satisfying $\dot{V} = 0$ $(x \neq 0)$ will be strictly positive; denoting this by V_{min}, it follows that the region

$$V(x) < V_{min}$$

is a domain of attraction.

11.7 Absolute Stability Criteria

The term 'absolute stability' (or hyperstability) is applied to certain types of nonlinear system which are known to be stable regardless of the exact form of the nonlinearity provided that its gain always lies within a given range. For example, consider the feedback system of Figure 11.1 .

Suppose the nonlinear function $f(u)$ satisfies $f(0) = 0$ and

$$\alpha < \frac{f(u)}{u} < \beta$$

for all $u \neq 0$, where α and β are real constants. Then, various conditions can be shown to be sufficient for global asymptotic stability, using Lyapunov functions or some other method; we shall not prove the results here but simply state them, for which purpose it is convenient to introduce the concept of 'positive-realness'.

A rational transfer function is said to be positive-real[4] if it has all the following properties: no poles in the right half-plane; any poles on the imaginary axis are simple and have real positive residues; the function has a non-negative real part at every point, apart from poles, on the imaginary axis. These properties are easily translated into conditions on the Nyquist plot of the function, enabling the stability criteria to be given a graphical interpretation. Assuming that $g(s)$ is strictly proper, we then have the following results.

The Circle Criterion. If the rational function

$$\frac{1+\beta g(s)}{1+\alpha g(s)}$$

is positive-real, then the nonlinear feedback system is globally asymptotically stable[3]. The graphical interpretation of this result depends on the signs of α and β.

For simplicity, suppose that $\beta > \alpha > 0$ and that $g(s)$ is the transfer function of an asymptotically stable system. Then, the circle criterion requires that the Nyquist plot of $g(j\omega)$ should not intersect or encircle the disc on $[-\alpha^{-1},-\beta^{-1}]$ as diameter.

It is clearly a generalisation of the Nyquist criterion for linear systems and is of theoretical importance but is usually rather conservative in practice.

The Popov Criterion. If $\alpha = 0$ and there exists a positive constant q such that the rational function

$$(1+qs)g(s) + \beta^{-1}$$

is positive-real, then the system is globally asymptotically stable[3]. Also, provided that the nonlinear function $f(u)$ is single-valued, this result can be extended to the case where q is negative. Like the circle criterion, it is only a sufficient (and not necessary) condition but is less conservative because of the freedom to choose the value of q. The graphical interpretation makes use of the so-called 'modified polar plot' of $\omega \mathrm{Im}\, g(j\omega)$ against $\mathrm{Re}\, g(j\omega)$; this plot is simply required to lie entirely to the right of a straight line (of slope q^{-1}) passing through the point $-\beta^{-1}$.

Results of this kind, employing the positive-realness concept, play an important part in nonlinear system theory and also have practical applications. A linear system with a positive-real transfer function is said to be 'passive' and this concept is of importance in the theory of *model-reference adaptive systems*[4], where the control system parameters are made to vary in a way which depends on the error between the outputs of the system and the model, so as to make the system's behaviour approach that of the model, Figure 11.3.

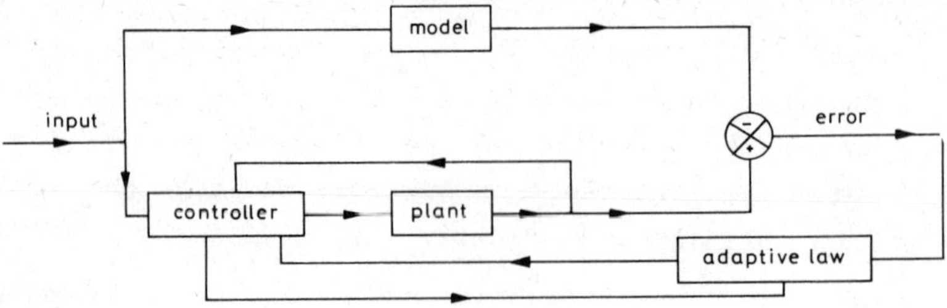

Fig. 11.3

Such systems are essentially nonlinear and it is necessary to design the adaptation procedure in such a way that global stability is ensured; in some cases, this can be done by utilising the properties of certain types of Lyapunov function which are known to exist for passive systems.

Another area of application for absolute stability criteria is in the *stability of interconnected systems*. The idea is that, if each of the subsystems is 'sufficiently stable' and the interconnections are 'weak enough', then the total system will be stable. In order to obtain quantitative conditions it is therefore necessary to have a measure of how stable the subsystems are and how strong are the connections between them. Such measures can be derived, for example, in terms of the properties of Lyapunov functions defined for the various subsystems; alternatively, an 'effective gain' may be defined for each subsystem, lower gain corresponding essentially to greater stability. Criteria can then be obtained which give sufficient conditions for global asymptotic stability of the whole system, the essential requirement being that the interactions should be sufficiently weak relative to the stability of the subsystems[5].

References

1. ATHERTON, D.P. : 'Nonlinear Control Engineering', (Van Nostrand Reinhold, 1975).

2. GELB, A. and W.E. VANDERVELDE : 'Multiple-input Describing Functions and Nonlinear System Design', McGraw-Hill, 1968).

3. WILLEMS, J.L. : 'Stability Theory of Dynamical Systems' (Nelson, 1970).

4. LANDAU, Y.D. : 'Adaptive Control - The Model Reference Approach', (Marcel Dekker, 1979).

5. COOK, P.A. : 'On the stability of interconnected systems', Int. J. Control, 20, (1974), pp. 407-415.

NONLINEAR SYSTEMS - Problems

P.11.1 Consider the feedback system of Fig. 11.1, with the nonlinearity being an ideal relay,

$$f(u) = \begin{cases} 1, & u > 0 \\ -1, & u < 0 \end{cases}$$

and $$g(s) = \frac{1}{s(s+1)(s+4)}$$

Show that a stable limit cycle is predicted, and estimate its amplitude and period of oscillation, using the describing function method.

P.11.2 For the system described by the state-space equations

$$\dot{x}_1 = x_2$$
$$\dot{x}_2 = -x_1 - x_2 + x_1 x_2$$

find the largest domain of attraction obtainable by using the Lyapunov function

$$V = x_1^2 + x_2^2$$

Chapter 12

Some DDC system design procedures

Dr. J. B. KNOWLES

Synopsis

As a result of the enormous impact of microprocessors, electronic engineers, with sometimes only a cursory background in control theory, are being involved in direct-digital-control (D.D.C.) system design. There appears to be a real need for an easily understood and simply implemented comprehensive design technique for single-input d.d.c. systems. The proposed design technique provides, first of all, a simple calculation that ensures that the data sampling rate is consistent with the control system's accuracy specification or the fatigue life of its actuators. Pulsed transfer-function design for a plant controller is based on two simple rules and a few standard frequency-response curves, which are easily computed once and for all time. Structural resonances are eliminated by digital notch filters, the pole-zero locations of which are directly related to the frequency and bandwidth of an oscillatory mode; this is exactly as with analogue networks. In addition a computationally simple formula gives an upper bound on the amplitude of the control error (deviation) component due to multiplicative rounding effects in the digital computer; this thereby enables the selection of a suitable machine wordlength or machine. A distinct advantage of the proposed design technique is that its implementation does not necessarily involve a complex computer-aided-design facility.

12.1 Introduction

As a result of the enormous impact of microprocessors, electronic engineers with sometimes only a cursory background in control theory are being involved in direct digital control (ddc) system design. There is now a real need[1] for an easily understood and simply implemented design technique for single-input DDC systems which is the objective of this tutorial paper. The proposed design procedure is shown as a flow chart in Figure 12.1, and it contains appropriate references to sections of text that treat the more important issues in detail. It is hoped that this diagram will provide a systematic approach to DDC system design, as well as an appreciation for the organisation of the text.

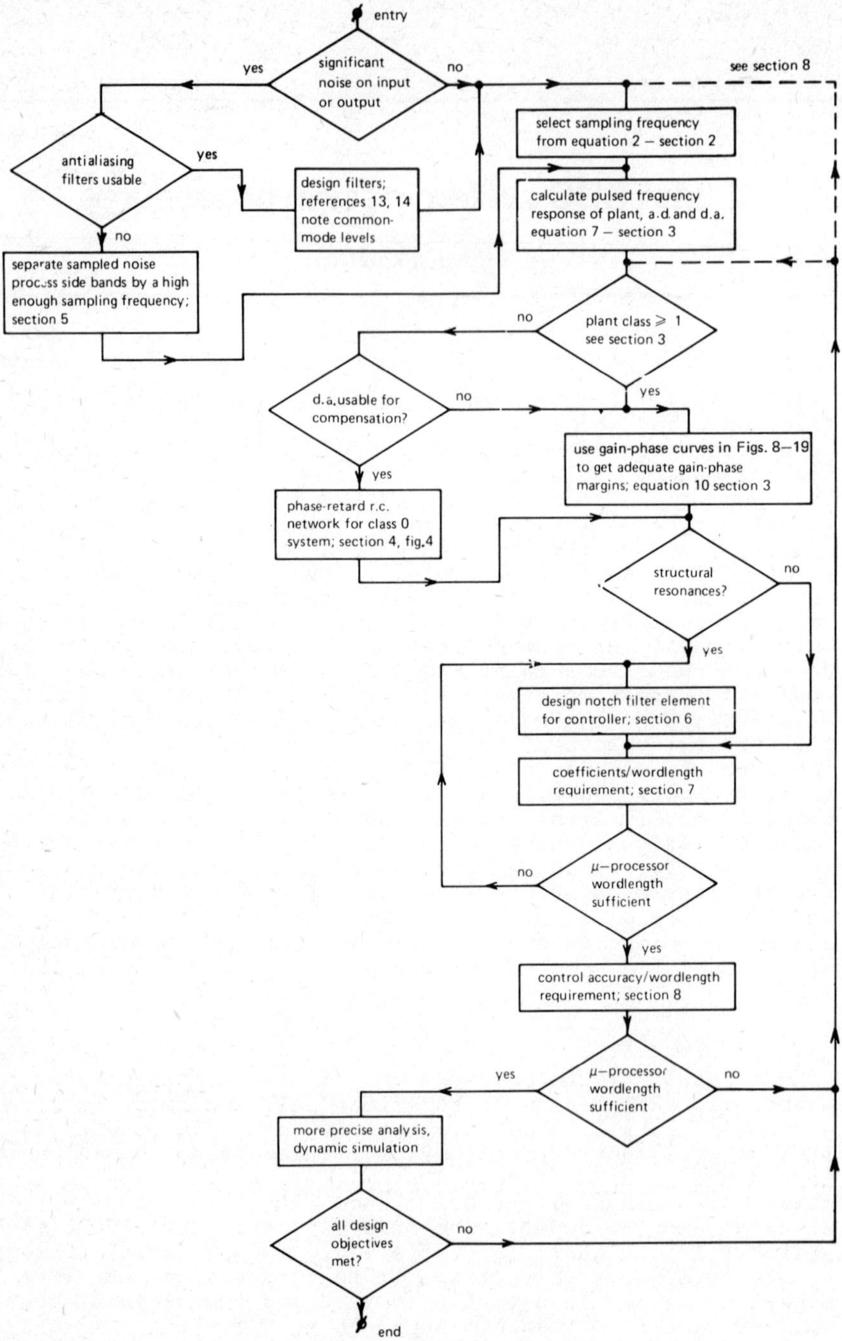

Fig. 12.1 d.d.c. design scheme

The experienced designer will notice the absence of such topics as:

(i) 'Bumpless transition' criteria during changes from automatic to manual control.
(ii) Provision for timed sequential operations to cope with fault conditions.

These aspects have been deliberately omitted because they are considered to be separate programming issues once control in the normal operating regime has been achieved.

Digital realisations of conventional three-term controllers have the advantages of: wide applicability, theoretical simplicity and ease of on-line tuning. However, the resulting closed-loop performance is generally inferior to that obtainable with other algorithms of similar numerical complexity. The graphical compensation procedure[2,17] described in Section 12.3 copes with the design of digital three-term controllers and these high performance units with equal ease. Also the technique is readily exploited by virtue of its simple calculations (amenable even to slide-rule treatment). Furthermore, it is shown to result in an 'even-tempered' closed-loop response for all input signals; unlike DDC systems synthesised by time-domain procedures.

Compared to analogue controllers, digital controllers offer distinct advantages in terms of: data transmission, interconnection, auxiliary data processing capabilities, fault tolerance, tamper resistance, etc.. However, a digital controller must evidently provide a control performance at least as good as that of the analogue controller it replaces. In this respect, it is suggested that sampling and wordlength effects are designed to be normally negligible relative to the control accuracy specification. When this objective is frustrated by computer performance constraints these degenerate effects can be evaluated from formulae given in the text.

12.2 Choice of Sampling Frequency

The first step in design of a ddc system is the selection of an appropriate sampling rate (T). Distortion in the form of spectral side-bands centred on integral multiples of the sampling frequency (1/T) is inherently produced by the periodic sampling of information, and the perfect (impractical) recovery of the signal requires the complete elimination of these harmonics[3]. A suitable practical choice of sampling frequency limits the distortion (or aliasing) by imposing a large enough frequency separation between the side-bands and the original unsampled signal spectrum for the low-pass plant elements to effect an adequate attenuation. Where comprehensive plant records for an existing analogue control scheme are available, the sampling period for a replacement ddc system is sometimes decided on the basis of requiring a 'small change' in the time dependent error or deviation signal during this interval. In the author's opinion, engineering judgements of this form are valuable only as confirmation of a fundamentally based and experimentally confirmed analysis.

For the sinusoidal input:

$$x(t) = A \sin(wt) \tag{12.1}$$

Knowles and Edwards[4] show that the average power of the control error component due to imperfect side-band attenuation is bounded for a unity-feedback, single-input ddc system by:

$$\overline{\varepsilon_R^2} \leq A^2 g^2 w_s^{-(2R+2)} M_p^2 w^2 |G(jw)|^{-2} \tag{12.2}$$

where $G(s)$ denotes the transfer function of the plant excluding the zero-order hold and:

$$G(s) = \left.\frac{g}{s^R}\right|_{s\to\infty} \quad ; \quad M_p = \text{Closed-loop peak magnification}$$

$$w_s = 2\pi/T \tag{12.3}$$

Equation (12.2) reveals that the high frequency components of the continuous input spectrum generate relatively the greatest distortion and loss of control accuracy. Thus, a simple design criterion for selecting the sampling frequency is to ensure that the right-hand-side of equation (12.2) represents an admissible loss of control accuracy for the largest permissible amplitude sinusoid at the highest likely signal frequency. This calculation is not only easily performed, but it is independent of subsequent stabilisation calculations. However, if computer loading demands the lowest possible consistent sampling frequency, it is then necessary to follow an interactive procedure which involves the digitally compensated system and the formula[4]:

$$\overline{\varepsilon_R^2} = \pi^{-3} g^2 w_s^{-2R} \int_{-\infty}^{\infty} |H^*(jw)/P^*(jw)|^2 \sin^2(wT/2)\Phi_x(w)dw \tag{12.4}$$

where $H^*(jw)$ defines the overall pulsed frequency response of the stabilised closed-loop system, $\Phi_x(w)$ is the input signal power spectrum, and $P^*(jw)$ denotes the pulsed transfer function of the plant and zero-order hold.

12.3 Frequency Domain Compensation Method

For most practical applications, necessary and sufficient conditions for a ddc system to be stable are that the open-loop pulsed frequency response:

$$\chi K(z)P(z)\Big|_{z=\exp(j2\pi fT)} \triangleq \chi K^*(jf)P^*(jf) \tag{12.5}$$

does not encircle the (-1, 0) point. Furthermore, in many cases the polar diagram of $\chi K^*(jf)P^*(jf)$ is such that an adequately damped time domain response is produced when its Gain and Phase Margins are greater than about 12dB and 50° respectively. It is evidently important to confirm that these stability margins are maintained over the full-range of plant parameter variations or uncertainties. Denoting the Laplace transfer function of the plant, A/D, D/A and data smoothing elements by $P(s)$ then the corresponding pulsed frequency response is obtained from references 3 and 5 as:

$$P^*(jf) = \frac{1}{T}\sum_{-\infty}^{\infty} P(j2\pi f - j2\pi n f_s) \tag{12.6}$$

Data sampling frequencies selected according to the criterion in Section 12.2 allow the above expression to be closely approximated by:

$$P^*(jf) \doteq \frac{1}{T}P(s)\Big|_{s=j2\pi f} \tag{12.7}$$

over the range of frequencies that determine closed-loop system stability. If the data smoothing element consists of a zero-order hold, and the combined gain of the A/D and D/A converters is χ_T, the required pulsed frequency response $P^*(jf)$ is readily calculable from:

$$P^*(jf) = \chi_T \exp(-j\pi fT) . \left| \frac{\sin(\pi fT)}{\pi fT} \right| . G(s) \Big|_{s=j2\pi f} \qquad (12.8)$$

For ease of nomenclature in the following discussion, the number of pure integrations in the plant transfer function $G(s)$ is termed the 'class' of the system. In general, it is recommended that one stage of phase-lead compensation is included for each pure integration in the plant. Thus, for a class 1 system, the discrete compensator takes the form:

$$\chi K(z) = \chi \left(\frac{z-a_1}{z-a_2}\right) \quad \text{with} \quad 0 \leq a_2 < a_1 < 1 \qquad (12.9)$$

where the pole at $z = a_2$ is necessary for physical realisability. Observe that digital realisations of P+I and P+D controllers can be represented in the form of equation (12.9). Graphs like Figures 12.2 and 12.3 (references 2,17), which show the gain and phase of $[\exp(j2\pi fT)-c]$ to base of normalised frequency (fT) for a limited number of real and complex pairs of values for c are the basis of the proposed design technique. The value of c corresponds to a controller zero or pole. As with analogue networks, increasing amounts of phase-lead are seen to be associated with increasingly severe rising gain-frequency characteristics. If the maximum phase-lead per stage is much above the recommended value of 55°, the rising gain-frequency response of the discrete controller causes the locus $\chi K^*(jf)P^*(jf)$ to bulge close to the (-1, 0) point which aggravates transient oscillations.

Pursuing a maximum bandwidth control system design, the measured or calculated values of $P^*(jf)$ are first used to determine† the frequency (f_B) for which:

$$\underline{/P^*(jf_B)} + 55^\circ \times \text{Number of Phase-lead Stages} = -130^\circ \qquad (12.10)$$

†A Bode or Nyquist diagram is helpful in this respect.

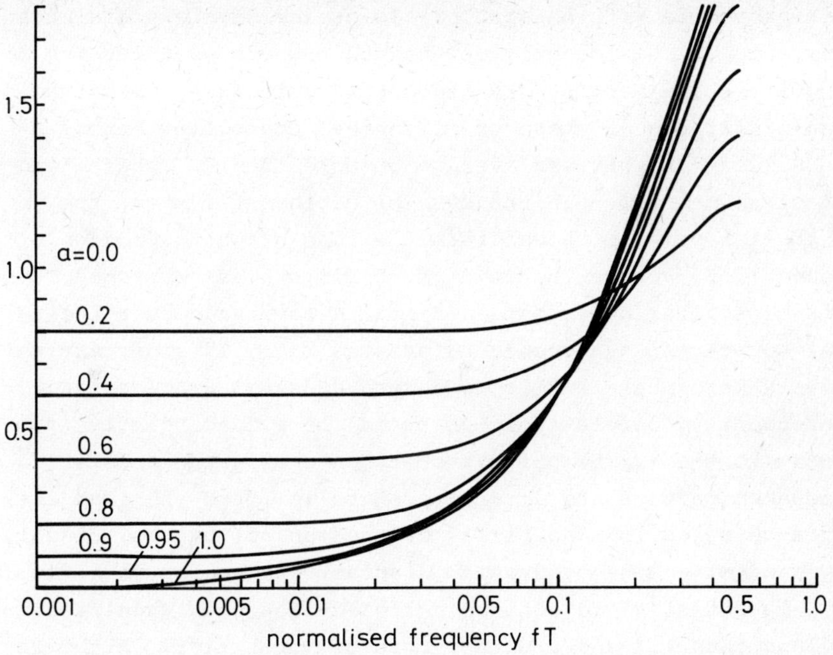

Fig. 12.2 Gain/normalised-frequency characteristics, $|\exp(j2\pi fT)-a|$

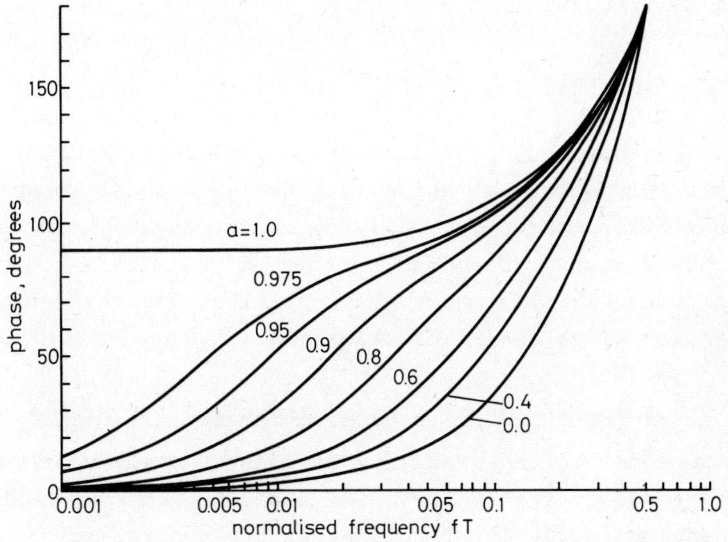

Fig. 12.3 Phase/normalised-frequency characteristics, $\angle\{\exp(j2\pi fT)-a\}$

This procedure with an appropriate choice of the scalar gain constant (χ) is clearly orientated towards establishing an adequate Phase Margin. By means of Figure 12.3, which shows $\underline{/\exp(j2\pi fT) - a}$ to base of fT, real controller zero(s) are selected to give on average no more than 70° phase-lead per stage at f_BT. In the case of a class 1 system, the controller pole is chosen from the same graph to produce around $15\text{-}20^{o}$ phase-lag at f_BT. For a class 2 system, complex controller poles have a definite advantage over their real counterparts, because references 2 and 17 show that relatively less phase-lag is generated at servo frequencies. Consequently, complex pole pairs may be placed relatively closer to the (1, 0) point thereby enhancing the zero-frequency gain of the discrete controller, and the load disturbance rejection capability of the control system. Finally, by means of a Bode or Nyquist diagram, the scalar gain factor (χ) is set so as to achieve a closed-loop peak magnification of less than 1.3 and a Gain-Margin of about 12dB. Although this maximum bandwidth design procedure may not be ideal for all applications it is nevertheless adaptable. For example, the pulsed transfer function of a conventional P+I controller may be written as:

$$KD(z) = (K_P+K_I)\left(\frac{z-a}{z-1}\right) \quad ; \quad a = \frac{K_P}{K_P+K_I} \qquad (12.11)$$

so that the gain and phase curves may be applied to select suitable values for 'a' and K_P+K_I, which uniquely specify the controller. Again, the design curves may be used to match the frequency response function of a digital controller to that of an existing analogue controller for the range of frequencies affecting plant stabilisation.

12.4 The Compensation of Class 0 (Regulator) Systems

The performance and stability of class 0 systems are generally improved by the inclusion of phase-retard compensators, whose analogue transfer function has the general form:

$$K(s) = \chi\left(\frac{1+s\alpha\tau}{1+s\tau}\right) \qquad (12.12)$$

where the attenuation constant α is less than unity but greater than 1/12 for practical applications. A/D converters[6] are frequently time-division multiplexed between several controllers, but each loop normally has exclusive use of its own D/A converter which can serve useful dual roles. A series resistor and capacitor connected across the feedback resistor of a D/A realises equation (12.12) and it is proposed for the phase-retard compensation of class 0 ddc systems. Apart from the benefits accruing from savings in computing time, the use of analogue networks in this instance is superior to digital because experience shows that an appreciable phase lag then occurs over a markedly narrower range of frequencies for the same attenuation constant. The importance of this fact will become apparent after the following description of phase-retard network design.

The first step in the compensation process is to determine the frequency ω_B rads/sec for which the phase of the frequency response function $P^*(j\omega)$ is -130°. If the Gain and Phase Margins of the uncompensated regulator are satisfactory, but the problem is to reduce the zero-frequency control error to ε_{DC}, then set

$$10/\alpha\tau = \omega_B \tag{12.13}$$

and

$$\alpha = \left(\frac{\varepsilon_{DC}}{1 - \varepsilon_{DC}}\right) P^*(0) \qquad \chi = 1/\alpha \tag{12.14a}$$

With these parameters the gain and phase of $K(s)$ for frequencies above ω_B are effectively unity and zero respectively. Hence, inclusion of the network increases the zero-frequency gain of the open-loop response without altering the original satisfactory stability margins. Alternatively, if the zero-frequency control accuracy is acceptable, but improvements in stability margins are desired, then select parameters according to equation (12.13) and

$$\chi = 1 \qquad \alpha = 1/|P^*(j\omega_B)| \quad \text{for a } 50^\circ \text{ Phase Margin} \tag{12.14b}$$

With these values, the gain and phase of $K(s)$ for frequencies above ω_B are α and zero respectively. Hence, inclusion of the network improves the Gain Margin by the factor $20 \log_{10} (1/\alpha)$, without altering the original satisfactory open-loop zero-frequency gain. Both the above design techniques for phase-retard networks are based on sacrificing phase in a relatively unimportant portion of the Nyquist diagram for gain in a separate region that markedly influences control accuracy or stability. As the frequency range over which the network produces an appreciable phase-lag widens, this simple design procedure becomes evidently complicated by the progressive overlap of these regions of gain and phase changes.

12.5 Noisy Input or Output Signals

The operational amplifier associated with a D/A converter can also be used for additional signal conditioning. Sometimes the input or output signal of a control system is heavily contaminated by relatively wide-bandwidth noise. If it is impractical to low-pass filter these signals to prevent aliasing[13,14], then an irrevocable loss of control performance can only be prevented by employing a sampling frequency which largely separates the side-bands of the sampled noise process. Unnecessary actuator operations, and therefore wear and tear, are sometimes avoided in this situation by including an averaging calculation in the digital compensator. In these circumstances, the side-band attenuation achieved by the plant is rendered less effective if the output sampling rate of the controller is reduced to an integral sub-multiple of its input rate. As an alternative solution to this noisy signal problem, it is proposed that a suitable capacitor is placed in parallel with the feedback resistor of the D/A converter.

12.6 Structural Resonances and Digital Notch Networks

A servo-motor is coupled to its load by inherently resilient steel drive shafts, and the combination possesses very selective resonances because the normal dissipative forces are engineered to be relatively small. In high power equipment (radio telescopes, gun mountings, etc.) the frequency of such structural resonances can be less than three times the required servo-bandwidth, and consequently their presence markedly complicates the compensation procedure. Viewed on a Nyquist diagram, the resonances cause the locus to loop-out again from the origin to encircle the (-1, 0) point. Analogue controllers obtain system stability under these conditions by processing the nodal error signal with tuned notch filters (e.g. Bridged-Tee) which block the excitation of each oscillatory mode. For a resonance peak at ω_0 rad/sec., the same effect is provided by the following digital filter:

$$N(z) = K_N \frac{(z-z_0)(z-z_0^*)}{(z-r_0 z_0)(z-r_0 z_0^*)} \qquad (12.15)$$

where

$$z_0 = \exp(j\omega_0 T)$$
$$z_0^* = \exp(-j\omega_0 T) \qquad 0 \le r < 1 \qquad (12.16)$$

and unity gain at zero-frequency is achieved by setting:

$$K_N = \left| \frac{1-r_0 z_0}{1-z_0} \right|^2 \qquad (12.17)$$

In order to null a mechanical resonance, its bandwidth and centre-frequency must both be matched by the digital filter. As may be appreciated from the pole-zero pattern of the filter, no larger notch bandwidth than necessary should be used in order to minimise the phase-lag incurred at servo-frequencies. It remains therefore to relate the notch bandwidth (B_{NO}) to the parameter r_0.

The gain-frequency response function of the proposed digital filter is given by:

$$|N^*(j\omega)| = K_N \left| \frac{(\exp(j\omega T)-z_0)(\exp(j\omega T)-z_0^*)}{(\exp(j\omega T)-r_0 z_0)(\exp(j\omega T)-r_0 z_0^*)} \right| \qquad (12.18)$$

Defining the frequency deviation variable:

$$\delta = \omega - \omega_0 \qquad (12.19)$$

then a power series expansion of exponential functions of δ yields the first order approximation:

$$|N^*(j\delta)| = \frac{K_{NO}}{\sqrt{1 + (1-r_0/T\delta)^2}} \qquad \text{for } \delta T << 1 \qquad (12.20)$$

where:

$$K_{NO} = K_N \left| \frac{1 - \exp(-j2\omega_0 T)}{1 - r_0 \exp(-j2\omega_0 T)} \right| \qquad (12.21)$$

Equation (12.20) evidently has complete similarity with the gain-frequency characteristic of an analogue notch filter in the vicinity of its anti-resonance. Accordingly, its bandwidth is defined as the frequency increment about ω_0 within which the attenuation is greater than 3dB:

$$B_{NO} = \pm\{(1 - r_0)/T\} \qquad (12.22)$$

Calculated or measured plant frequency responses may therefore be used in conjunction with equations (12.16) and (12.22) to specify suitable values for ω_0 and r_0 in equation (12.15), and the resulting filter is incorporated as an algebraic factor in the pulse function of the discrete controller.

12.7 Coefficient Quantisation in a Discrete Controller

The coefficients in a pulse transfer function usually require rounding in order to be accommodated in the finite wordlength format of a microprocessor or minicomputer. Because generous stability margins are normally employed, coefficient rounding is normally unimportant as regards the practical realisation of phase-lead or phase-retard pulse transfer functions. However, it is well-known that increasingly severe digital filter specifications (cut-off rate, bandwidth, etc.) accentuate computer wordlength requirements[8,9]. Hence it is generally prudent to examine the wordlength necessary to counter acute structural resonances by the proposed form of digital notch filter.

By writing equation (12.15) as:

$$N(z) = K_N \frac{z^2-az+1}{z^2 - rbz+r^2} \tag{12.23}$$

coefficient rounding on $\{a, rb, r^2\}$ is seen to modify both the centre-frequency and the bandwidth of a notch filter design. Ideally with no rounding the coefficients in equation (12.23) are specified by:

$$a = a_0; \quad (rb) = (r_0 b_0); \quad a_0 = b_0 = 2\cos(\omega_0 T); \quad (r^2) = (r_0^2) \tag{12.24}$$

and defining for the practical situation the variables

$$a = 2\cos(\omega_1 T); \quad b = 2\cos(\omega_2 T) \tag{12.25}$$

$$\delta = \omega - \omega_1 \quad ; \quad \varepsilon = \omega_1 - \omega_2 \tag{12.26}$$

one obtains the gain-frequency response function of the realisations as:

$$|N^*(j\omega)| = K_N \left|\frac{\exp(j\delta)-1}{\exp(j\delta+j\varepsilon)-r}\right| \left|\frac{1 - \exp[-j(\omega+\omega_1)T]}{1 - r\exp[-j(\omega+\omega_2)T]}\right| \tag{12.27}$$

Thus the centre-frequency of the filter is not at ω_1, and the shift due to the finite computer wordlength is evaluated directly from equation (12.25) as:

$$\delta\omega_0 = \frac{1}{T}[\cos^{-1}(a/2) - \cos^{-1}(a_0/2)] \text{ rad/sec} \tag{12.28}$$

It is interesting to note that the change in centre-frequency cannot be obtained by differentiating equation (12.25) because a second-order approximation is required if the coefficient a_0 approaches -2. In the vicinity of the anti-resonance and with fine quantisation:

$$\delta T \ll 1 \quad \text{and} \quad \varepsilon T \ll 1 \tag{12.29}$$

and under these conditions a power series expansion of the exponential terms in equation (12.27) yields:

$$|N^*(j\delta)| \doteq \frac{K_N'}{\sqrt{(1+\varepsilon/\delta)^2 + ((1-r)/\delta T)^2}} \tag{12.30}$$

where

$$K_N' = K_N \left| \frac{1 - \exp(-j2\omega_1 T)}{1 - r\exp(-j2\omega_2 T)} \right| \tag{12.31}$$

Thus the bandwidth (B_N) of the realisation satisfies:

$$(1 + \frac{\varepsilon}{B_N})^2 + (\frac{1-r}{B_N T})^2 = 2 \tag{12.32}$$

or

$$B_N^2 = 2\varepsilon B_N + \varepsilon^2 + (\frac{1-r}{T})^2 \tag{12.33}$$

For small enough perturbations about the ideal situation defined in equation (12.24), it follows that:

$$\left|\delta B_{NO}\right| \leq \left|\frac{\partial B_N}{\partial \varepsilon}\right|_0 \left|\delta\varepsilon\right| + \left|\frac{\partial B_N}{\partial r}\right|_0 \left|\delta r\right| \tag{12.34}$$

From equations (12.24), (12.26) and (12.33) one obtains:

$$\begin{gathered} \delta\varepsilon \leq 2\,\delta\omega_0 \;; \qquad \delta r \leq q/2r_0 \\ \left|\frac{\partial B_N}{\partial \varepsilon}\right|_0 = 1 \;; \qquad \left|\frac{\partial B_N}{\partial r}\right|_0 = 1/T \end{gathered} \tag{12.35}$$

where the width of quantisation is defined by:

$$q = 2^{-(\text{wordlength} - 1)} \tag{12.36}$$

Hence, the change in the notch bandwidth of the filter due to coefficient rounding is bounded by:

$$|\delta B_{NO}| \leq 2|\delta\omega_0| + \frac{q}{2T(1-B_{NO}T)} \tag{12.37}$$

As a design example, consider the realisation of a notch

filter with: centre frequency 12 rad/sec, bandwidth ±1 rad/sec and a sampling period of 0.1 sec. Equation (12.24) defines the ideal numerator coefficient as:

$$a_0 = 0.724716 \qquad (12.38)$$

With an 8-bit machine format, the actual numerator coefficient realised is:

$$a = 0.718750 \quad (0.1011100) \qquad (12.39)$$

so that from equation (12.28):

$$\delta\omega_0 = 0.032 \text{ rad/sec} \qquad (12.40)$$

and the change in filter bandwidth evaluates from equation (12.37) as:

$$\delta B_{NO} \leq 0.11 \text{ rad/sec} \qquad (12.41)$$

Thus, in practical terms, an 8-bit microprocessor is about sufficient for realising this particular specification.

12.8 Arithmetic Roundoff-Noise in a Discrete Controller

Arithmetic multiplications implemented by a digital computer are subject to rounding errors due to its finite wordlength. As a result, random noise is generated within the closed loop and the control accuracy is degenerated. In extreme cases, the control system can even become grossly unstable. It is therefore important to relate the loss of control accuracy to the wordlength of the discrete controller. The analysis in Reference 10 provides an easily calculable upper-bound for the amplitude of the noise process present on the error signal due to arithmetic rounding in a discrete controller. With a directly programmed compensator for example, this upper-bound is given by:

$$|\varepsilon_Q| \leq \left(\frac{\text{Number of Significant Multiplications in } \chi K(z)}{2\chi|\Sigma \text{ Numerator Coefficients}|} \right) q \qquad (12.42)$$

where a 'significant' multiplication does *not* involve zero or a positive integral power of 2 including 2^0. During a normally iterative design procedure, equation (12.42) is valuable as a means of comparing the multiplicative rounding error generated by various possible controllers. In preliminary calculations, the pessimism* of this upper-bound is of little consequence. However a more accurate estimate of performance degeneration due to multiplicative rounding errors can be economically important for the finalised design when it is realised on a bit-conscious microprocessor. For this purpose, the analysis in References 11 and 12 is recommended because one straightforward calculation quantifies the loss of control accuracy as a function of controller wordlength.

As described earlier, the sampling frequency of a DDC system must be high enough to prevent an unacceptable loss of control accuracy or fatigue damage to plant actuators. However, strangely enough, it is possible in practical terms for the sampling frequency to be made too high. Reference to Figure 12.3 shows that by increasing the sampling frequency, the phase-lead required at a particular frequency can only be sustained by moving the compensator zero(s) closer to unity. To preserve load disturbance rejection, the zero-frequency gain of the controller $(\chi K(z)|_{z=0})$ must be maintained by then moving the pole(s) also closer to unity. As a result, the algebraic sum of the compensator's numerator coefficients is reduced, and multiplicative round-off noise on the error signal is increased according to equation (12.42). Thus the choice of an unnecessarily high sampling frequency can entail unjustified expense in meeting long wordlength requirements.

12.9 Multirate and Subrate Controllers

A subrate digital controller has its output sampler operating an integral number of times slower than its input. In

---

*The formula over-estimated by about 2 bits in the example considered in Reference 10.

a multi-rate controller, the output sampler operates an integral number of times faster than its input. Section 12.5 describes how subrate systems prejudice ripple spectrum attenuation, and in practice excessive actuator wear seems best prevented for noisy input signals by the use of a single rate controller and a smoothing capacitor across the D/A unit. It is now pertinent to question if multi-rate systems afford any advantages over their more easily designed single-rate counter-parts.

As a result of the higher pulse rate exciting the plant, Kranc[15] contends that the ripple performance of a multi-rate system is superior to that of a single rate system, even though both cases have the same input sampling rate. However, it should be noted that the input sampler of the multi-rate sampler still generates spectral side-bands centred on multiples of $\pm\omega_0$, and that these are not eliminated in further modulation by the faster output sampler. Hence, it is by no means self-evident that increasing the output sampling rate alone in a digital compensator effects an improvement in ripple attenuation. To clarify this issue and other aspects of performance, a comparison of single and multi-rate unity feedback systems is implemented in Reference 16 for the plant and zero-order hold transfer function:

$$P(s) = \frac{e^{-sT}(1-e^{-sT})}{s^3(s+1)} \tag{12.43}$$

with computer sampling periods of:

$$\left.\begin{aligned} T_{in} &= T_{out} = 0.1 \text{ sec} \quad \text{Single-rate system} \\ T_{in} &= 3T_{out} = 0.1 \text{ sec} \quad \text{Multi-rate system} \end{aligned}\right) \tag{12.44}$$

Both closed-loop systems are compensated to have virtually identical frequency and transient response characteristics. For two typical input spectra, calculations show that the output ripple power for the multi-rate system is at least a decade greater than that for the single rate system. An intuitive explanation of this result may be based on the fact that in both systems the compensator and zero-order hold are attempting to predict some function of the error over the input sampling period. As each prediction inevitably incurs error, a multi-rate system makes more errors per unit time and thus gives the larger ripple signal. It may be argued that multi-rate controllers give potentially the faster transient response. However, the conclusion reached in Reference 16 is that improvements in transient response leads to a further degradation in ripple performance.

12.10 Comparison of Time Domain Synthesis and Frequency Domain Compensation Techniques

A convenient starting point is Figure 12.4, which shows the step and ramp responses of a time domain synthesised dead-beat system whose continuous part consists of a D/A converter and the plant:

$$P(s) = \frac{10}{s(s+1)^2} \tag{12.45}$$

with a sampling period of 1 second. Though the ramp response of this example (which is taken from Reference 3) is good, the 100% overshoot in the step response is unacceptable for most purposes. It is emphasised that this 'highly-tuned' behaviour is typical of ddc systems which have been synthesised in the time domain by the dead-beat or minimum variance techniques[3,5]. The example is therefore considered unbiased, and a useful vehicle for comparing time domain methods with the proposed frequency domain compensation procedure. For the Class 1 system in equation (12.45) the algorithm in Section 12.3 specifies a series compensating element of the form:

$$\chi K(z) = \chi \frac{(z-a)}{(z-b)} \tag{12.46}$$

whose maximum phase-lead must not exceed about 55^0. In order to achieve an adequate phase-margin ($\simeq 50^o$), the frequency (f_B) at which the phase lag of $P^*(j\omega)$ is 180° indicates the maximum bandwidth likely to be achieved. It is readily seen from equation (12.46) that in this case:

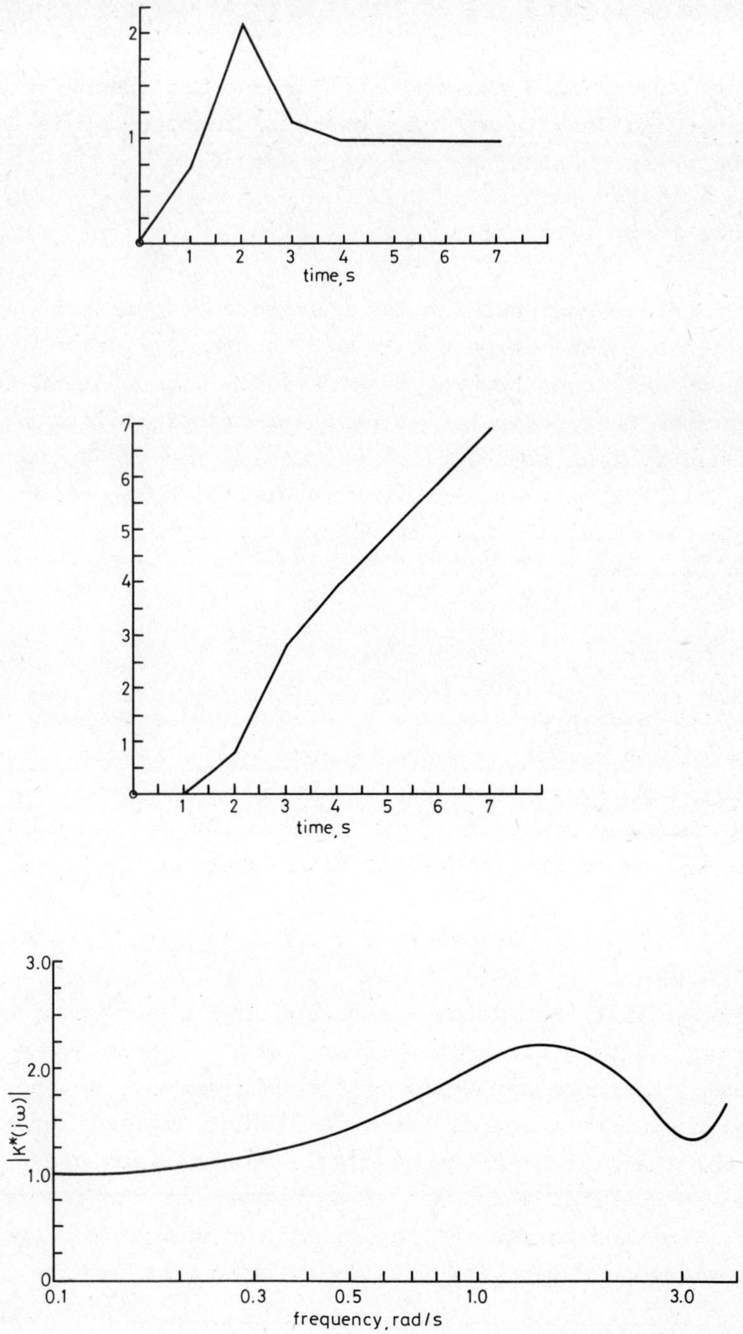

Figure 12.4 Response Characteristics of Dead-Beat System

$$f_B T \simeq 0.16 \qquad (12.47)$$

However, Figure 12.3 shows that the phase difference between numerator and denominator of a discrete compensator is then well below the maximum which occurs around:

$$fT = 0.03 \qquad (12.48)$$

More effective compensation can therefore be realised by decreasing the sampling period to 0.25 seconds*. Calculations of the pulsed frequency response $P^*(j\omega)$ via equation (12.8) differ from those computed using the z-transformation in gain and phase by less than 0.01%; so vindicating the approximation. As $f_B T$ now equals 0.035, a digital compensator which provides 55° lead at this frequency is selected by means of Figure 12.3 as:

$$\chi K(z) = \chi \frac{(z-0.9)}{(z-0.4)} \qquad (12.49)$$

which with a gain constant (χ) of 0.4 realises a gain-margin of approximately 10dB. Figure 12.5 shows that the step and ramp function responses are now both adequately damped, and the additional control error due to multiplicative rounding errors is derived from equation (12.42) as:

$$|\varepsilon_Q| \leq 37.5q \qquad (12.50)$$

which indicates that double-precision arithmetic would be necessary on an 8-bit μ-processor. Apart from an 'even-tempered' response and wordlength requirements, the proposed design procedure also provides: a simpler digital compensator, the stability margins available to cope with plant variations or uncertainties, and an assessment of the rms ripple error. For the purpose of designing a practical ddc system it provides necessary information so evidently lacking in the

*Note the rms ripple error at 1 rad/s even with a sampling period of 1s is satisfactory; equation (12.2) evaluates as 1.3×10^{-3}.

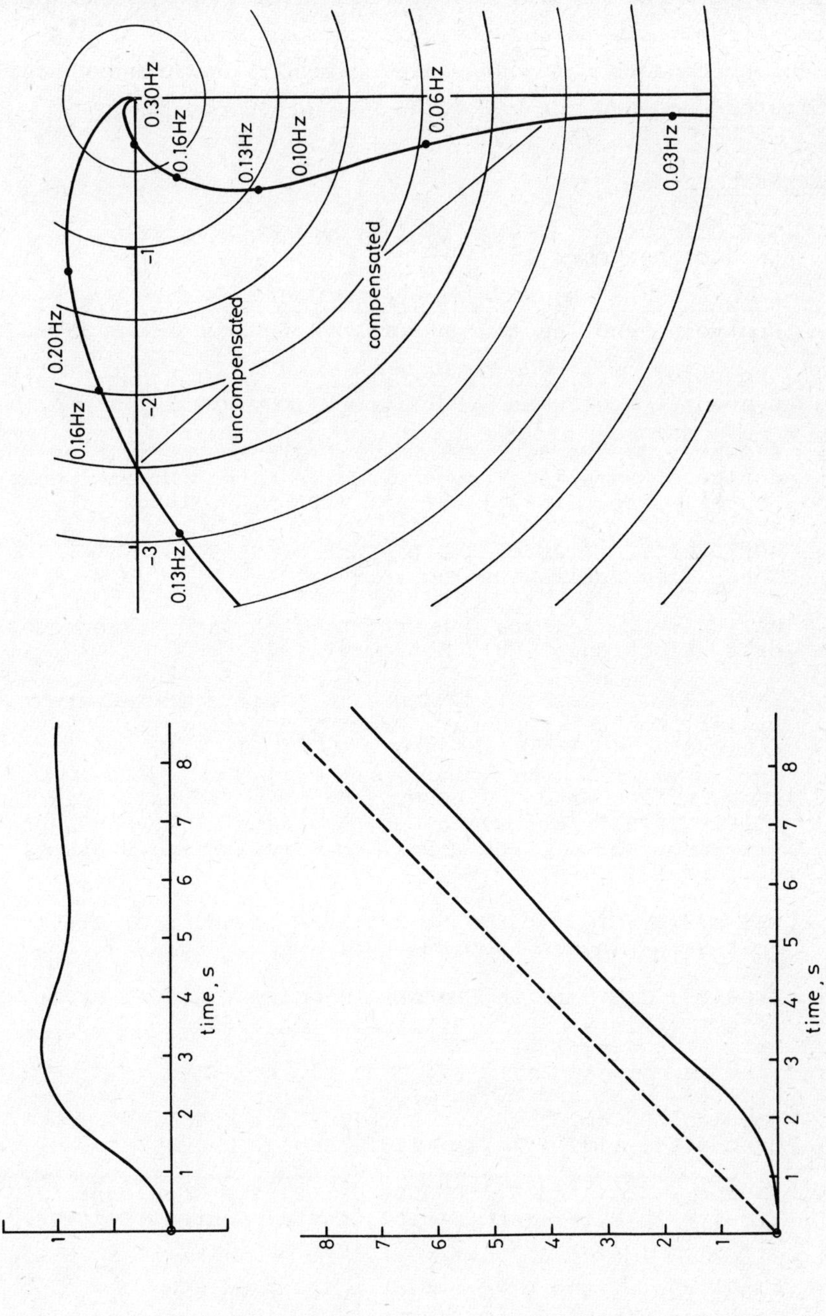

Fig. 12.5 Response Characteristics of Frequency Domain Compensated System.

time domain techniques. Furthermore, the design rules are demonstrably simple and the calculations involve about half a dozen standard gain and phase curves[2,17], which are easily computed once and for all time.

REFERENCES

1. IEE Coloquium: Design of Discrete Controllers, London, December 1977.

2. KNOWLES ,J.B. : A Contribution to Direct Digital Control, PhD Thesis, University of Manchester, 1962.

3. RAGAZZINI, J.R. and G. FRANKLIN : Sampled Data Control Systems. McGraw-Hill Publishers Inc., 1958.

4. KNOWLES, J.B. and R. EDWARDS : Ripple Performance and Choice of Sampling Frequency for a Direct Digital Control System. Proc. IEE, 113,(11), (1966), p. 1885.

5. JURY, E.I. : Sampled-data Control Systems. John Wiley Publishers Inc., 1958.

6. MATTERA, L. : Data Converters Latch onto Microprocessors. Electronics, 81, September 1977.

7. KNOWLES, J.B. and R. EDWARDS : Aspects of Subrate Digital Control Systems. Proc. IEE, 113, (11), 1966, p. 1893.

8. AGARWAL, R.C. and C.S. BURRUS : New Recursive Digital Filter Structures have a Very Low Sensitivity and Round off Error Noise. IEE Trans. on Circuits and Systems, 22, No. 12, 1975, p. 921.

9. WEINSTEIN, C.J. : Quantization Effects in Digital Filters. Lincoln Laboratory MIT Rept. 468, Nov. 1969.

10. KNOWLES, J.B. and R. EDWARDS : Computational Error Effects in a Direct Digital Control System. Automatica, 4, 1966, 7.

11. KNOWLES, J.B. and R. EDWARDS : Effect of a Finite Wordlength Computer in a Sampled-Data Feedback System. Proc. IEE, 112, 1965, p. 1197.

12. KNOWLES, J.B. and R. EDWARDS : Finite Wordlength Effects in Multi-rate Direct Digital Control Systems. Proc. IEE, 112, 1965, p. 2377.

13. WONG, Y.T. and W.E. OTT, (of Burr-Brown Inc.) : Function Circuits - Design and Applications. Published by McGraw Hill, 1976.

14. STEWART, R.M. : Statistical Design and Evaluation of Filters for the Restoration of Sampled Data. Proc. IRE, 44, 1956, 253.

15. KRANC, G.M. : Compensation of an Error-Sampled System by a Multi-rate Controller. Trans AIEE, 76, 1957, p. 149.

16. KNOWLES, J.B. and R. EDWARDS : Critical Comparison of Multi-rate and Single-rate DDC System Performance. JACC Computational Methods Session II.

17. KNOWLES, J.B. : Comprehensive, yet Computationally Simple Direct Digital Control-system Design Technique. Proc. IEE, 125, December 1978, p. 1383.

Chapter 13

Robust controller design

Professor E. J. DAVISON

Synopsis

An overview of some recent results in the controller design of multivariable linear time-invariant systems, in which uncertainty/perturbation in plant behaviour occurs is made. In particular, the *servomechanism problem* is considered for various cases varying from the extreme case when the plant is known and fixed, to the opposite extreme when the plant is completely unknown and is allowed to be perturbed. In addition, the chapter includes the case of time lag systems, decentralized systems and the control of systems subject to sensor/actuator failure. The emphasis throughout is to characterize structural results for the servomechanism problem, e.g. to determine the existence of a solution to the problem and to determine the controller structure required to solve the problem under different conditions. The chapter concludes with a proposed design method for obtaining realistic robust controller design.

13.1 Introduction

This chapter shall attempt to look at some aspects of controller synthesis for the multivariable servomechanism problem when the plant to be controlled is subject to uncertainty - in this case, a controller is to be found so that satisfactory regulation/tracking occurs in spite of the fact that the parameters of the plant may be allowed to vary (by arbitrary large amounts), subject only to the condition that the resultant controlled perturbed system remains stable. This type of controller design will be called *robust controller design* in this chapter. It is to be noted that other definitions of "robust controller design" may be found in the literature.

The topic of robust control is an extremely active area of research at present, and so it is impossible to include all results/approaches obtained to date. Some recent literature dealing with various aspects of robust control design by using a number of different approaches may be found, e.g. in Sain[17]; Spang, Gerhart[19]. No attempt shall be made in this short chapter to survey all such approaches. Instead, the chapter shall concentrate only on some of the structural results recently obtained re the servomechanism problem (in the time domain) and discuss one possible approach for obtaining effective "robust controller design".

The chapter is divided as follows: part 2 describes the general problem, part 3 describes some structural results obtained for the servomechanism problem under different conditions, part 4 describes some extensions of the previous results and part 5 describes a proposed design method for obtaining realistic robust controller design.

13.2 General Problem to be Considered

A plant has inputs u which can be manipulated, outputs y which are to be regulated, outputs y_m which can be measured, disturbances ω which, in general, cannot be measured, and reference input signals y_{ref} which it is desired that the outputs "track". The general servomechanism problem consists of finding a controller (either centralized or decentralized) which has inputs y_m, y_{ref} and outputs u for the plant so that:

1. The resultant closed loop system is stable	Stability
2. Asymptotic tracking occurs, independent of any disturbances appearing in the system	Regulation
3. Fast response occurs 4. "Low interaction" occurs	Transient behaviour
5. (1)-(4) occur independent of any variations in the plant parameters of the system (including changes of plant model order)	Robust behaviour
6. "Non-explosive" behaviour occurs in the closed-loop system under sensor/actuator failure	Integrity
7. Under sensor/actuator failure, the remaining parts of the controlled system still satisfactorily track/regulate	Reliability

In addition, depending on the particular physical problem many other constraints may be imposed on the problem statement, e.g. damping factor constraint, controller gain constraint, etc.

13.3 The Servomechanism Problem - Structural Results

13.3.1 Problem description

The *plant* to be controlled is assumed to be described by the following linear time-invariant model:

$$\begin{aligned} \dot{x} &= Ax + Bu + E\omega \\ y &= Cx + Du + F\omega \qquad\qquad (13.1) \\ y_m &= C_m x + D_m u + F_m \omega \quad , \quad e = y - y_{ref} \end{aligned}$$

where $x \varepsilon R^n$ is the state, $u \varepsilon R^m$ is the output, $y \varepsilon R^r$ is the output to be regulated, $y_m \varepsilon R^{r_m}$ are the measurable outputs of the system $\omega \varepsilon R^{\Omega}$ are the disturbances in the system, $y_{ref} \varepsilon R^r$ are the reference inputs and $e \varepsilon R^r$ is the error in the system.

It is assumed that the *disturbances* ω arise from the following system:

$$\begin{aligned} \dot{\eta}_1 &= A_1 \eta_1 \quad , \qquad \eta_1 \varepsilon R^{n_1} \qquad\qquad (13.2) \\ \omega &= C_1 \eta_1 \end{aligned}$$

and that the *reference inputs* y_{ref} arise from the system:

$$\begin{aligned} \dot{\eta}_2 &= A_2 \eta_2 \ , \quad \eta_2 \varepsilon R^{n_2} \\ \sigma &= C_2 \eta_2 \qquad\qquad (13.3) \\ y_{ref} &= G\sigma \end{aligned}$$

It is assumed for nontrivality that $sp(A_1) \subset C^+$, $sp(A_2) \subset C^+$ where $sp(\cdot)$ denotes the eigenvalues of $(\cdot)$ and C^+ denotes the closed right half complex plane. It is also assumed with no loss of generality, that (C_1, A_1), (C_2, A_2) are observable and that rank C=r, rank B=m, rank $C_m = r_m$, rank $\begin{bmatrix} E \\ F \end{bmatrix}$ = rank $C_1 = \Omega$, rank G = rank $C_2 = \dim(\sigma)$.

The following specific types of servomechanism problems are now posed for the plant (13.1):

I. Feedforward Servomechanism Problem:

Find a controller for (13.1) (with input y_m, y_{ref} and output u) so that:

(a) The resultant closed-loop system is stable.

(b) Asymptotic tracking occurs, i.e.

$$\lim_{t\to\infty} e(t) = 0, \quad \forall\ x(0)\varepsilon R^n,\ \forall\ \eta_1(0)\varepsilon R^{n_1},\ \forall\ \eta_2(0)\varepsilon R^{n_2},$$

and for all controller initial conditions.

II. Perfect Control Problem

Find a controller for (13.1) so that (a), (b) above hold and in addition for any $\delta > 0$:

(c) $\left|\left|\int_0^\infty e^T Q e d\tau\right|\right| < \delta, \quad \forall$ bounded $x(0)\varepsilon R^n$, $\eta_1(0)\varepsilon R^{n_1}$, $\eta_2(0)\varepsilon R^{n_2}$

and all bounded controller initial conditions.

III. Feedforward Control Problem with Feedback Gain Perturbations

Assume the following controller is applied to (13.1):

$$u = K_0\eta + K_1 y_m + K_2 y_{ref}, \quad \dot{\eta} = \Lambda_0\eta + \Lambda_1 y_m + \Lambda_2 y_{ref}$$

then it's desired to find a controller for (13.1), so that (a), (b) above hold and in addition:

(d) Condition (b) holds for any variations in K_0, K_1, K_2 which do not cause the resultant system to become unstable.

IV. Weak Robust Servomechanism Problem

Find a controller for (13.1) so that conditions (a), (b), (d) above hold and in addition:

(e) Condition (b) holds for any variations in the elements of (A,B) which do not cause the resultant system to become unstable.

V. Robust Servomechanism Problem

Find a controller for (13.1) so that conditions (a), (b), (d) above hold and in addition:

(f) Condition (b) holds for any variations in the plant model (including changes of model order) which do not cause the resultant system to become unstable.

VI. Strong Robust Servomechanism Problem

Find a controller for (13.1) so that conditions (a), (b), (d), (f) hold and in addition:

(g) Approximate error regulation occurs for any variations in the parameters of the controller used, the approximation becoming arbitrarily close to the exact solution as the controller parameter variation becomes smaller.

VII. Perfect Control Robust Servomechanism Problem

Find a controller for (13.1) so that conditions (a), (b), (c), (d) and (f) hold.

VIII. Robust Decentralized Servomechanism Problem

Find a controller for (13.1) so that conditions (a), (b), (d), (f) hold and in addition:

(h) The resultant controller is decentralized[8].

IX. Sequentially Reliable Servomechanism Problem

Find a decentralized controller for (13.1) so that conditions (a), (b), (d), (f), (h) hold and in addition:

(i) If a failure occurs in the ith control agent's controller, ($1 \leq i \leq \nu$), conditions (a), (b), (d), (f), (h) still apply to the remaining part of the system consisting of control agents $1,2,\ldots,i-1$.

X. Robust Servomechanism Problem for Plants with Time Lags

Given that the plant (13.1) is now modelled by

$$\dot{x} = A_0 x + A_1 x(t-\Delta) + Bu + E\omega$$
$$y_m = C_m^0 x + C_m^1 x(t-\Delta) + D_m u + F_m \omega \qquad (13.4)$$
$$y = C_0 x + C_1 x(t-\Delta) + Du + F\omega$$

find a controller so that conditions (a), (b), (d), (f) hold.

13.3.2 Existence Results

The following definitions are required in the development to follow.

Definition

Consider the system (13.1), (13.2), (13.3); then the system (C,A,B,D,E,F) is said to be *steady-state invertible with respect to the class of inputs* $(C_1,A_1;C_2,A_2,G)$ if and only if the following two conditions hold:

$$\text{rank} \begin{pmatrix} I_{n_1} \otimes A - A_1^T \otimes I_n & I_{n_1} \otimes B \\ I_{n_1} \otimes C & I & I_{n_1} \otimes D \end{pmatrix}$$

$$= \text{rank} \begin{pmatrix} I_{n1} \otimes A - A_1^T \otimes I_n & I_{n_1} \otimes B & (I_{n_1} \otimes E)\underline{C}_1 \\ I_{n_1} \otimes C & I_{n_1} \otimes D & (I_{n_1} \otimes F)\underline{C}_1 \end{pmatrix}$$

$$\text{rank} \begin{pmatrix} I_{n_2} \otimes A - A_2^T \otimes I_n & I_{n_2} \otimes B \\ I_{n_2} \otimes C & I_{n_2} \otimes D \end{pmatrix}$$

$$= \text{rank} \begin{pmatrix} I_{n_2} \otimes A - A_2^T \otimes I_n & I_{n_2} \otimes B & O \\ I_{n_2} \otimes C & I_{n_2} \otimes D & (I_{n_2} \otimes G)\underline{C}_2 \end{pmatrix}$$

Definition

Consider the system (13.1), (13.2) and assume that A, A_1 have disjoint eigenvalues (note that if (C_m,A,B) is stabilizable and detectable, this can always be achieved by applying static output feedback from y_m); then the system

$(C,A,E,F:C_m,F_m)$ is said to be *output-detectable with respect to the class of inputs* (C_1,A_1) if and only if:

$$\text{rank}\begin{bmatrix} F_mC_1+C_mT \\ (F_mC_1+C_mT)A_1 \\ \vdots \\ (F_mC_1+C_mT)A_1^{n_1-1} \end{bmatrix} = \text{rank}\begin{bmatrix} FC_1+CT \\ F_mC_1+C_mT \\ (F_mC_1+C_mT)A_1 \\ \vdots \\ (F_mC_1+C_mT)A_1^{n_1-1} \end{bmatrix}$$

where T is the solution of $AT-A_1T = -EC_1$.

Definition

Let the minimal polynomial of A_1, A_2 be denoted by $\Lambda_1(s)$, $\Lambda_2(s)$ respectively, and let the least common multiple of $\Lambda_1(s)$, $\Lambda_2(s)$ be given by $\Lambda(s)$. Let the zeros of $\Lambda(s)$ (multiplicities included) be given by $\lambda_1,\lambda_2,\ldots,\lambda_p$. Then $\lambda_1,\lambda_2,\ldots,\lambda_p$ *are called the disturbance/reference signal poles of (13.2), (13.3).*

The following results are obtained re the existence of a solution to the servomechanism problems defined previously.

I. Feedforward Servomechanism Problem

Theorem 1[5,6]

There exists a solution to the feedforward servomechanism problem for (13.1) if and only if the following conditions all hold:

(i) (C_m,A,B) is stabilizable and detectable.

(ii) (C,A,B,E,F) is steady-state invertible with respect to the class of inputs $(C_1,A_1;C_2,A_2G)$.

(iii) $(C,A,E,F;C_m,F_m)$ is output detectable with respect to the class of input (C_1,A_1).

The following is a sufficient condition which guarantees that theorem 1 always holds.

Corollary 1

There exists a solution to the feedforward servomechanism problem for (13.1) if the following conditions all hold:

(i) $(C_m, A, B,)$ is stabilizable and detectable.

(ii) $m \geq r$.

(iii) the transmission zeros[2] of (C, A, B, D) do not coincide with λ_i, $i = 1, 2, \ldots, p$.

(iv) either $y_m \equiv y$ or $\omega \equiv 0$ or $r_m \geq \Omega$ and the transmission zeros of $(C_m . A, E, F_m)$ do not coincide with any eigenvalues of A_1.

II. Perfect Control Problem

Theorem 2[10,13]

There exists a solution to the perfect control problem for (13.1) if and only if the following conditions all hold:

(i) (C, A, B) is stabilizable and detectable.

(ii) $m \geq r$.

(iii) the transmission zeros of (C, A, B, D) are contained in the open left part of the complex plane (i.e. the plant is minimum phase).

(iv) the states x, η_1, η_2 are available for measurement.

The following result follows directly from theorem 2 and shows the importance of using to full advantage the total number of manipulated inputs which a plant may have.

Theorem 3 (Control Inequality Principle)[10,13]

Given the system (13.1), then:

(i) If $m > r$, there exists a solution to the perfect control problem for "almost all" systems (13.1).

(ii) If $m = r$, there may or may not exist a solution to the perfect control problem depending on whether the plant is minimum or non-minimum phase.

(iii) If $m < r$, there never exists a solution to the perfect control problem.

III. Feedforward Control Problem with Feedback Gain Perturbations

Theorem 4[15]

There exists a solution to the feedforward control problem with feedback gain perturbations if and only if the following conditions all hold:

(i) there exists a solution to the feedforward control problem, i.e. theorem 1 holds.

(ii) the output y_m must contain either "partial errors" or "equivalent errors" or both (i.e. the output y_m must contain some of the outputs y).

IV. Weak Robust Servomechanism Problem

Theorem 5[7,3]

There exists a solution to the weak robust servomechanism problem for (13.1) if and only if the following conditions all hold:

(i) (C_m, A, B) is stabilizable and detectable

(ii) $m \geq r$

(iii) the transmission zeros of (C,A,B,D) do not coincide with λ_i, $i = 1,2,\ldots,p$

(iv) $\operatorname{rank} C_m = \operatorname{rank} \begin{bmatrix} C_m \\ C \end{bmatrix}$

V. Robust Servomechism Problem

Theorem 6[7,3]

There exists a solution to the robust servomechanism problem for (13.1) if and only if the following conditions all hold:

(i) (C_m, A, B) is stabilizable and detectable

(ii) $m \geq r$

(iii) the transmission zeros of (C,A,B,D) do not coincide with λ_i, $i = 1,2,\ldots,p$

(iv) the outputs of y are physically measurable.

VI. Strong Robust Servomechanism Problem

Theorem 7[3]

There exists a solution to the strong robust servomechanism problem for (13.1) if and only if the following conditions both hold:

(i) there exists a solution to the robust servomechanism problem, i.e. theorem 6 holds.

(ii) the class of disturbance and reference input signals is bounded, i.e. the disturbance, reference signals must have the property that $Re(\lambda_i) = 0$, $i = 1,2,\ldots,p$ and that $\lambda_1, \lambda_2, \ldots, \lambda_p$ are all distinct.

VII. Perfect Control Robust Servomechanism Problem

Theorem 8 (Robust Control Limitation Theorem)[10,13]

There exists no solution to the perfect control robust servomechanism problem.

VIII. Robust Decentralized Servomechanism Problem

Assume that the plant (13.1) consists of ν control agents modelled by:

$$\dot{x} = Ax + \sum_{i=1}^{\nu} B_i u_i + E\omega$$

$$\left.\begin{aligned} y_m^i &= C_m^i x + D_m^i u_i + F_i^m \omega \\ y_i &= C_i x + D_i u_i + F_i \omega \end{aligned}\right\} \quad i = 1,2,\ldots,\nu \qquad (13.5)$$

$$e_i \triangleq y_i - y_i^{ref}$$

where $y_i \in R^{r_i}$ is the output to be regulated of the ith control station, y_m^i are the measurable outputs of the ith control station, u_i are the control inputs of the ith control station and e_i is the error in the ith control station. It is assumed that the disturbance signals ω satisfy (13.2) and that the reference signals $y_{ref} \triangleq \begin{pmatrix} y_1^{ref} \\ \vdots \\ y_\nu^{ref} \end{pmatrix}$ satisfy (13.3).

Definition

Let

$$C \triangleq \begin{bmatrix} C_1 \\ C_2 \\ \vdots \\ C_\nu \end{bmatrix}, \quad D \triangleq \text{block diag}(D_1, D_2, \ldots, D\ ),$$

$$B \triangleq (B_1, B_2, \ldots, B_\nu), \quad C_m \triangleq \begin{pmatrix} C_m^1 \\ C_m^2 \\ \vdots \\ C_m^\nu \end{pmatrix}$$

and let

$$C_m^* \triangleq \left| \begin{matrix} \begin{pmatrix} C_1^m & 0 & 0 & \cdots & 0 \\ 0 & I_{r_1} & 0 & \cdots & 0 \end{pmatrix} \\ \begin{pmatrix} C_2^m & 0 & 0 & \cdots & 0 \\ 0 & 0 & I_{r_2} & \cdots & 0 \end{pmatrix} \\ \vdots \\ \begin{pmatrix} C_\nu^m & 0 & 0 & \cdots & 0 \\ 0 & 0 & 0 & \cdots & I_{r_\nu} \end{pmatrix} \end{matrix} \right|$$

Theorem 9[8]

There exists a solution to the robust decentralized servomechanism problem for (13.5) if and only if the following conditions all hold:

(i) (C_m, A, B) has no unstable decentralized fixed modes.

(ii) The decentralized fixed modes of the p systems

$$\left\{ C_m^*, \begin{bmatrix} A & 0 \\ C & \lambda_j I \end{bmatrix}, \begin{bmatrix} B \\ D \end{bmatrix} \right\}, \quad j = 1,2,\ldots,p$$

do not contain λ_j, $j = 1,2,\ldots,p$, respectively.

(iii) y_m^i contains the output y_i, $i = 1,2,\ldots,\nu$, respectively.

IX. Sequentially Reliable Robust Servomechanism Problem

Definition

Let $\lambda_{p+1} \triangleq 0$

Theorem 10[14]

Assume that the plant (13.1) is open loop asymptotically stable; then there exists a solution to the robust decentralized servomechanism problem for (13.5) with sequential reliability if and only if the following conditions all hold:

(i) there exists a solution to the robust centralized servomechanism problem for (13.1) with control agent 1 for the class of disturbance/reference signal poles

$\lambda_1, \lambda_2, \ldots, \lambda_p, \lambda_{p+1}$.

(ii) there exists a solution to the robust decentralized servomechanism problem for (13.1) with control agents 1, 2 for the class of disturbance/reference signal poles $\lambda_1, \lambda_2, \ldots, \lambda_p, \lambda_{p+1}$.

⋮

(v) there exists a solution to the robust decentralized servomechanism problem for (13.1) with control agents $1, 2, \ldots, \nu$ for the class of disturbance/reference signal poles $\lambda_1, \lambda_2, \ldots, \lambda_p, \lambda_{p+1}$.

X. Robust Servomechanism Problem for Plants with Time Lags

Definition

Given (13.4), let A be defined as follows: given the system $\dot{x} = A_0 x + A_1 x(t-\Delta) + Bu$, it may be written in the form:

$$\dot{x}_t = Ax_t + Bu$$

where $x_t \in M_2$ and where:

$$Ax_t = \begin{cases} \dfrac{dx_t}{d\theta} & , \quad \theta \in [-\Delta, 0) \\ A_0 x_t(0) + A_i x_t(-\Delta) & , \quad \theta = 0 \end{cases}$$

$$Bu = \begin{cases} 0 & , \quad \theta \in [-\Delta, 0) \\ Bu & , \quad \theta = 0 \end{cases}$$

The following result is obtained:

Theorem 11[16]

There exists a solution to the robust servomechanism problem for (13.4) if and only if the following conditions all hold:

(i) $\operatorname{rank}[A_0 + A_1 e^{-\lambda\Delta} - \lambda I, B] = n, \quad \forall\, \lambda \in sp(A)$ with $\operatorname{Re}(\lambda) \geq 0$

(ii)

$$\text{rank}\begin{bmatrix} A_0+A_1e^{-\lambda\Delta}-\lambda I \\ C_m^0+C_m^1e^{-\lambda\Delta} \end{bmatrix} = n, \ \forall \ \lambda \in sp(A) \text{ with } Re(\lambda)\geq 0$$

(iii)

$$\text{rank}\begin{bmatrix} A_0+A_1e^{-\lambda_i\Delta}-\lambda_i I & B \\ C_0+C_1e^{-\lambda_i\Delta} & D \end{bmatrix} = n+r, \ i = 1,2,\ldots,p$$

(iv) y is measurable.

13.3.3 Robust Servomechanism Controller Structure

The following definition of a stabilizing compensator and servocompensator are required in the development to follow.

Definition (Stabilizing Compensator)

Given the stabilizable-detectable system (C_m,A,B,D_m) represented by

$$\dot{x} = Ax + Bu$$

$$y_m = C_m x + D_m u$$

a *stabilizing compensator*: $u = K_1\eta+K_2y_m$, $\dot{\eta} = \Lambda_1\eta+\Lambda_2y_m$ is defined to be a controller which will stabilize the resultant closed loop system such that "desired" transient behaviour occurs. It is not a unique device and may be designed by using a number of different techniques, e.g. observer theory, LQR theory, output pole assignment methods, modal design methods, frequency design methods, etc. The ability to achieve a "desired" transient behaviour depends on the location of the transmission zeros of (C_m,A,B,D), the fixed modes of (C_m,A,B) etc.

Consider now the system (13.1) and assume that a solution to the robust servomechanism problem exists, i.e. theorem 6 holds. The following definition is needed:

Definition

Given the disturbance/reference poles λ_i, $i = 1,2,\ldots,p$, the matrix $C \in R^{p\times p}$ and vector $\gamma \in R^p$ is defined by:

$$C \triangleq \begin{bmatrix} 0 & 1 & 0 & \cdots & 0 \\ 0 & 0 & 1 & \cdots & 0 \\ \vdots & \vdots & \vdots & & \vdots \\ -\delta_1 & -\delta_2 & -\delta_3 & \cdots & -\delta_p \end{bmatrix}, \quad \gamma \triangleq \begin{bmatrix} 0 \\ 0 \\ \vdots \\ 0 \\ 1 \end{bmatrix}$$

where the coefficients $_i$, $i = 1,2,\ldots,p$ are given by the coefficients of the polynomial $\prod_{i=1}^{p} (\lambda-\lambda_i)$, i.e.

$$\lambda^p + \delta_p\lambda^{p-1} + \ldots + \delta_2\lambda + \delta_1 \triangleq \prod_{i=1}^{p} (\lambda-\lambda_i)$$

The following compensator, called a *servo-compensator* is of fundamental importance in the design of robust controllers for (13.1):

<u>Definition</u> (Servo-Compensator)

Consider the system (13.1), (13.2), (13.3); then a *servo-compensator* for (13.1) is a controller with input $e \in R^r$ and output $\xi \in R^{rp}$ given by:

$$\dot{\xi} = \mathcal{C}^*\xi + \mathcal{B}^*e \qquad (13.7)$$

where

$$\mathcal{C}^* \triangleq \text{block diag}(\underbrace{\mathcal{C},\mathcal{C},\ldots,\mathcal{C}}_{r}), \quad \mathcal{B}^* \triangleq \text{block diag}(\underbrace{\gamma,\ ,\ldots,\gamma}_{r})$$

where $\mathcal{C}$, γ are given by (13.6). The servo-compensator is unique within the class of coordinate transformations and nonsingular input transformations.

<u>Robust Servomechanism Controller</u>

Consider the system (13.1); then any robust controller for (13.1) consists of the following structure:

$$u = K_0\hat{x} + K \qquad (13.8)$$

where $\xi \in R^{rp}$ is the output of the servo-compensator (13.7) and where $\hat{x}$ is the output of a *stabilizing compensator* $\mathcal{S}^*$,(with inputs y_{ref}, y_m, ξ, u) where $\mathcal{S}^*$, K_0, K are found to stabilize and give desired transient behaviour to the following augmented system:

$$\begin{pmatrix} \dot{x} \\ \dot{\xi} \end{pmatrix} = \begin{pmatrix} A & 0 \\ \mathcal{B}^{*}C & \mathcal{C}^{*} \end{pmatrix} \begin{pmatrix} x \\ \xi \end{pmatrix} + \begin{pmatrix} B \\ \mathcal{B}^{*}D \end{pmatrix} u \qquad (13.9)$$

$$\begin{pmatrix} y_m \\ \xi \end{pmatrix} = \begin{pmatrix} C_m & 0 \\ 0 & I \end{pmatrix} \begin{pmatrix} x \\ \xi \end{pmatrix} + \begin{pmatrix} D_m \\ 0 \end{pmatrix} u$$

where the fixed modes (if any) of $\left\{ \begin{pmatrix} C_m & 0 \\ 0 & I \end{pmatrix}, \begin{pmatrix} A & 0 \\ \mathcal{B}^{*}C & \mathcal{C}^{*} \end{pmatrix}, \begin{pmatrix} B \\ \mathcal{B}^{*}D \end{pmatrix} \right\}$

are equal to the fixed modes of (C_m, A, B).

In the above robust controller, the error input e must be obtained by using the physical outputs corresponding to y. It is to be noted that the robust controller always has order $\geq rp$.

13.3.4 Some Properties of the Robust Controller[3]

1. In the above robust controller, it is only necessary to know the disturbance/reference poles $\lambda_1, \lambda_2, \ldots, \lambda_p$, i.e. it is not necessary to to know $E, F, G, C_1, C_2, A_1, A_2$.
2. The robust controller is robust for *any changes* (i.e. not just small perturbations) in the plant parameters $(C, A, B, D; C_m, D_m; E; F; F_m)$, feedback gain parameters K_0, K, stabilizing compensator parameters of S^*, and for *any changes in the order* of the assumed mathematical model describing the plant, provided only that the resultant closed loop system remains stable.
3. A robust controller exists for "almost all" plants (13.1) provided that (i) $m \geq r$ and (ii) the outputs to be regulated can be physically measured; if either (i) or (ii) fail to hold then there never is a solution to the robust servomechanism problem.

13.3.5 Various Classes of Stabilizing Compensators

Various special classes of stabilizing compensators which can be used in the robust servomechanism controller S^* (13.8) are given as follows:

1. Observer-Stabilizing Compensator[12]: Assume that theorem 6 holds; then an observer stabilizing compensator S^* for (13.8) consists of the following device: $\hat{x}$ in (13.8) is the output of an observer of order $\hat{n}$ (either full order $\hat{n} < n$ or reduced order $\hat{n} \leq n-r$) with input y_m, u which estimates the state of the detectable system: $\dot{x} = Ax+Bu$, $y_m = C_m x+D_m u$, and where K_0, K in (13.8) are found, using standard methods, to stabilize the following stabilizable system:

$$\begin{pmatrix} \dot{x} \\ \dot{\xi} \end{pmatrix} = \begin{pmatrix} A & 0 \\ B^*C & C^* \end{pmatrix}\begin{pmatrix} x \\ \xi \end{pmatrix} + \begin{pmatrix} B \\ B^*D \end{pmatrix} u, \quad u = (K_0, K)\begin{pmatrix} x \\ \xi \end{pmatrix} \tag{13.10}$$

In case the plant (13.1) is open loop asymptotically stable and that $\mathrm{Re}(\lambda_i) = 0$, $i = 1,2,\ldots,p$, the following generalized three-term controller can be used for S^*.

2. Generalized Three-Term Controller: Consider the system (13.1); then the following compensator

$$u = K_0\xi + K_1\hat{y}_m + K_2\dot{\tilde{y}}_m \tag{13.11}$$

where $\hat{y}_m \triangleq y_m - D_m u$, $\dot{\tilde{y}}_m \triangleq \dot{\hat{y}}_m - C_m Bu$, where ξ is the output of the servo-compensator (13.7) is called a generalized three-term controller for (13.1). This compensator has the following property.

Theorem 12[12]

Consider the system (13.1) and assume that A is asymptotically stable, $\mathrm{Re}(\lambda_i) = 0$, $i = 1,2,\ldots,p$ and that theorem 6 holds; then there exists K_0, K_1, K_2 so that the resultant controlled system obtained by applying the generalized three term controller (13.11) to (13.1) is asymptotically stable.

Remark 1: In this case, arbitrary pole assignment cannot necessarily be obtained using (13.11) unlike the case of the observer stabilizing compensator. However, it follows from Davison, Wang (1975) that $(m+2r+rp-1)$ symmetric poles can be assigned to the closed-loop system for "almost all" (C_m, A, B) systems. This implies that if n is not too large, i.e. if

$n \leq (m+2r-1)$ that complete pole assignment can be carried out using (13.11) (even if A is unstable).

13.4 Extension of Previous Results

13.4.1 Control of Unknown Systems (Multivariable Tuning Regulators)

The previous results assumed that the plant model (13.1) is known, and based on a knowledge of (13.1), conditions for the existence of a robust controller and the construction of a controller were obtained. The question may now be asked - to what extent can one determine the existence of a solution and obtain a robust controller, when the plant model (13.1) is *completely unknown*? The following result shows that by carrying out some simple experiments on the plant (actual), the existence of a solution and the construction of a robust controller can be obtained. In this case, asymptotic regulation/tracking can be achieved, but the ability to achieve arbitrary fast speed of response, as was obtained when the plant is known, cannot now be necessarily carried out.

The following assumptions are made:

Assumptions

(i) The plant (actual) can be described by a linear time-invariant model and is open loop asymptotically stable.

(ii) The disturbance/tracking poles have the property that $\mathrm{Re}(\lambda_i) = 0$, $i = 1,2,\ldots,p$.

(iii) The outputs y can be measured.

The following definition is now made:

Definition

The *tracking steady-state gain parameters* of (13.1) are defined as follows:

$$T_i \triangleq D + C(\lambda_i I - A)^{-1}B\ , \quad i = 1,2,\ldots,p$$

Remarks

2. The tracking steady-state gain parameters of a system can be easily identified by sequentially applying m linearly

independent control signals to the plant (actual) and by measuring the steady-state values of the outputs y for each of the input signals. In this case, the input signals applied to the plant are of the same class as the class of disturbance/tracking signals for which the controller is designed.

3. It is to be emphasized that no other assumptions re (13.1), e.g. the order of the plant, minimum phase conditions, controllability-observability, etc., are required in this problem statement. This differs from the well known self-tuning regulator theory of Astrom (1973).

Existence Result

Assume that T_i, $i = 1,2,\ldots,p$ have been experimentally obtained for the plant. The following existence result can then be applied:

Theorem 13[9]

There exists a solution to the robust servomechanism problem for (13.1) if and only if

$$\text{rank } T_i = r \qquad i = 1,2,\ldots,p$$

Controller Structure

Assuming that T_i, $i = 1,2,\ldots,p$ has been determined and that theorem 13 holds, a robust controller can be obtained for the plant by carrying out a sequence of one-dimensional on line experiments on the plant with the servo-compensator (13.7) connected. This process of adjusting a single gain parameter is called "tuning the controller on line"; in this case, it can be guaranteed that the resultant closed-loop system always remains asymptotically stable. For further details see reference 9.

13.4.2 Multivariable Error Constants

In classical control, the position, velocity and acceleration errors can be used as an indication of "how well" a controller which is designed to give zero asymptotic error for a given class of signals, will behave when subject to a larger class of disturbance/reference signals than the design signals. The following generalization of these definitions can be used in the robust servomechanism problem to give an indication of "how well" the robust controller will behave when subject to a larger class of signals than the design signals. The following special case of error constants is obtained from reference 18.

Given the plant (13.1), assume that theorem 6 holds for the case of polynomial disturbance/reference signals (i.e. with $\lambda_i = 0$, $i = 1,2,\dots,p$), and assume the following controllers have been applied to (13.1) so that the resultant system is asymptotically stable.

Controller 0

$$u = K(y_{ref}-y) + \{K_1 + K_0(sI-\Lambda_0)^{-1}\Lambda_1\}\hat{y}_m, \quad \hat{y}_m \triangleq (y_m-D_m u)$$

Controller 1

$$u = \frac{K}{s}(y_{ref}-y) + \{K_1+K_0(sI-\Lambda_0)^{-1}\Lambda_1\}\hat{y}_m$$

Controller 2

$$u = \frac{K^1+K^2 s}{s^2}(y_{ref}-y) + \{K_1+K_0(sI-\Lambda_0)^{-1}\Lambda_1\}\hat{y}_m$$

Here, K/s, $K^1+K^2 s/s^2$ is a servo-compensator for (13.1) for the case of constant, ramp signals respectively, and $K_1+K_0(sI-\Lambda_0)^{-1}\Lambda_1$ is a stabilizing compensator for the system.

Let $e \triangleq y_{ref}-y$; then the following steady-state errors are obtained for the controlled system when reference signals $y_{ref} = \bar{y}t^i$, $i = 0,1,2$ where $\bar{y}$ is a constant, and disturbance signals $\omega = \bar{\omega}t^i$, $i = 0,1,2$ where $\bar{\omega}$ is a constant, are applied to the plant controlled by controllers 0,1,2 (see Tables 13.1, 13.2).

TABLE 13.1 Generalized Steady-State Errors for Various Reference Signals

	$y_{ref}=\bar{y}$	$y_{ref}=\bar{y}t$	$y_{ref}=\bar{y}t^2$
Controller 0	$\varepsilon_{t_0}\bar{y}$	∞	∞
Controller 1	0	$\varepsilon_{t_1}\bar{y}$	∞
Controller 2	0	0	$\varepsilon_{t_2}\bar{y}$

where:

$\varepsilon_{t_0} \triangleq \{I+\hat{C}[A+B(K_1-K_0\Lambda_0^{-1}\Lambda_1)C_m]^{-1}BK\}^{-1}$ Tracking Position Error Gain

$\varepsilon_{t_1} \triangleq \{\hat{C}[A+B(K_1-K_0\Lambda_0^{-1}\Lambda_1)C_m]^{-1}BK\}^{-1}$ Tracking Velocity Error Gain

$\varepsilon_{t_2} \triangleq 2\{\hat{C}[A+B(K_1-K_0\Lambda_0^{-1}\Lambda_1)C_m]^{-1}BK^1\}^{-1}$ Tracking Acceleration Error Gain

where:

$$\hat{C} \triangleq C + D(K_1-K_0\Delta_0^{-1}\Delta_1)C_m$$

TABLE 13.2 Generalized Steady-State Errors for Various Disturbance Signals

	$\omega=\bar{\omega}$	$\omega=\bar{\omega}t$	$\omega=\bar{\omega}t^2$
Controller 0	$\varepsilon_{d_0}\bar{\omega}$	∞	∞
Controller 1	0	$\varepsilon_{d_1}\bar{\omega}$	∞
Controller 2	0	0	$\varepsilon_{d_2}\bar{\omega}$

where

$\varepsilon_{d_0} \triangleq -\varepsilon_{t_0}\{\hat{C}[A+B(K_1-K_0\Lambda_0^{-1}\Lambda_1)C_m]^{-1}\hat{E}+\hat{F}\}$ Disturbance Position Error Gain

$\varepsilon_{d_1} \triangleq -\varepsilon_{t_1}\{\hat{C}[A+B(K_1-K_0\Lambda_0^{-1}\Lambda_1)C_m]^{-1}\hat{E}+\hat{F}\}$ Disturbance Velocity Error Gain

$\varepsilon_{d_2} \triangleq -\varepsilon_{t_2}\{\hat{C}[A+B(K_1-K_0\Lambda_0^{-1}\Lambda_1)C_m]^{-1}\hat{E}+\hat{F}\}$ Disturbance Acceleration Error Gain

where:

$$\hat{E} \triangleq E + B(K_1-K_0\Lambda_0^{-1}\Lambda_1)F_m$$

$$\hat{F} \triangleq F + D(K_1-K_0\Lambda_0^{-1}\Lambda_1)F_m$$

It is seen therefore that if a robust controller is designed to give asymptotic regulation for the class of constant signals say (Controller 1), it is clearly desirable to have the velocity error gains as "small as possible" (subject to other constraints in the controller design).

13.5 Robust Controller Design

For a given servo-compensator and stabilizing-compensator structure, the question of determining the "optimal" feedback gains and parameters of the stabilizing compensator has not been addressed. Generally it is desirable to determine these parameters so as to stabilize the overall closed-loop system, maximize speed of response in the system, minimize interaction between the reference signals and outputs to be regulated, subject to various constraints on the system such as integrity in the case of sensor/actuator failure, damping factor constraints, etc.

This is a topic of immense activity of which a number of different design approaches have been suggested. One such approach that can deal with constraints in a natural manner is to pose the problem as a parameter optimization problem, i.e. the gain parameters and stabilizing compensator parameters are found so as to minimize a performance index which reflects the speed of response and interaction of the closed loop system, subject to various constraints on the closed loop system. The following gives a brief review of such a design approach[11,12].

Problem Statement

Given the plant (13.1), assume λ_i, $i = 1,2,\ldots,p$ is given, that theorem 6 holds and that the stabilizing compensator

$$u = K_0 y_m + K_1 \eta + K\xi \ , \quad \dot{\eta} = \Lambda_0 \eta + \Lambda_1 y_m \qquad (13.12)$$

is to be used where $\dim(\eta)$ is specified; then the closed loop system is described by (13.1), (13.7) and (13.12).

Definition

Let $z \triangleq D\xi$ where ξ is the output of the servo-compensator (13.7) where

$$D \triangleq \text{block diag}\{\underbrace{(1,0,\ldots,0),(1,0,\ldots,0),\ldots,(1,0,\ldots,0)}_{r}\},$$

and let z_∞, u_∞ denote the steady-state value of z, u for a given disturbance/reference-input signal. Let $\Delta u \triangleq u-u_\infty$, $\Delta z \triangleq z-z_\infty$.

Performance Index

Let

$$J \triangleq E\int_0^\infty (\Delta z^T Q \Delta z + \mu \Delta u^T R \Delta u)\,d\tau, \quad Q > 0,\ R > 0,\ \mu > 0 \qquad (13.13)$$

where $\mu \to 0$, be the performance index for the system, where the expected value is taken with respect to all initial conditions $x(0)$, $\xi(0)$, $\eta(0)$, $\eta_1(0)$, $\eta_2(0)$ uniformly distributed on a unit sphere.

Then the following parameter optimization problem is to be solved:

Parameter Optimization Problem

$$\min_{K_0,K_1,K,\Lambda_0,\Lambda_1} J \quad \text{subject to various constraints imposed on the problem formulation}$$

For further details see references 11, 12.

13.5.1 Design Example

The following example gives an illustration of the proposed design procedure. The parameter optimization was carried out for the case $Q = I$, $R = I$, $\mu = 10^{-8}$.

Example 13.1 (Unstable batch reactor)

An unstable batch reactor (with open loop eigenvalues 1.99, 0.064, -5.057, -8.67) described by:

$$\dot{x}=\begin{pmatrix} 1.38 & -0.2077 & 6.715 & -5.676 \\ -0.5814 & -4.29 & 0 & 0.675 \\ 1.067 & 4.273 & -6.654 & 5.893 \\ 0.048 & 4.273 & 1.343 & -2.104 \end{pmatrix} x + \begin{pmatrix} 0 & 0 \\ 5.679 & 0 \\ 1.136 & -3.146 \\ 1.136 & 0 \end{pmatrix} u + E\omega$$

$$y = \begin{pmatrix} 1 & 0 & 1 & -1 \\ 0 & 1 & 0 & 1 \end{pmatrix} x$$

is to be regulated against constant disturbances such that the output tracks constant set-points y_{ref}, subject to the *integrity constraint* that the system's open loop unstable eigenvalues (i.e. $Re(\lambda) \leq 1.99\times1.1 = 2.19$) for a sensor failure in output y_1 or y_2.

In this case, since the system has transmission zeros (-1.192, -5.039) and no fixed modes, there exists a solution to the robust servomechanism problem from theorem 6. The following robust controller is obtained from (13.8)

$$u = K_0 y + K_1 \int_0^t (y-y_{ref})d\tau$$

and it is desired to determine K_0, K_1 so as to minimize J given by (13.13) subject to the above integrity constraint. The following result is obtained:

$$J_{opt} = 1.39 \times 10^{-2}$$

$$\text{closed loop eigenvalues} = (-1.19, -5.04, -125\pm j125, -168\pm j169)$$

$$(K_0, K)_{optimal} = \left(\begin{array}{cc|cc} 0.95 & -58.4 & 100 & -1.0\times10^4 \\ 79.3 & -0.72 & 1.0\times10^4 & -100 \end{array}\right)$$

In this case the constraints are satisfied; in particular the active constraint occurs when the output sensor for y_1 fails, and the resulting eigenvalues obtained in this case are given by (2.15, 0, -1.01, -8.66, -168$\pm$j169). It is to be noted that with this controller a "pole-zero" cancellation occurs with respect to the eigenvalues -1.19, -5.04 . Some typical responses of the closed-loop system are given in Figure 13.1 for the case of set-point change in y_{ref} and for the case of an arbitrary constant disturbance

$E = (1 \; 1 \; 1 \; 1)'$ for zero initial conditions. It is seen that excellent tracking and regulation occur.

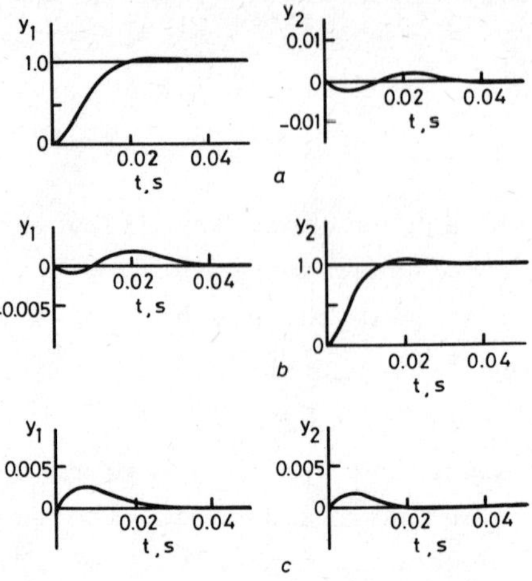

Fig. 13.1 Response of controlled system for example 1 using proposed design method.
(a) $y_{ref} = \begin{pmatrix} 1 \\ 0 \end{pmatrix}$, $\omega = 0$, (b) $y_{ref} = \begin{pmatrix} 0 \\ 1 \end{pmatrix}$, $\omega = 0$,
(c) $y_{ref} = \begin{pmatrix} 0 \\ 0 \end{pmatrix}$, $E\omega = (1 \; 1 \; 1 \; 1)$

13.6 Conclusions

A brief overview of some recent results obtained re the robust control servomechanism problem has been outlined in this chapter. In particular, structural results for the following classes of servomechanism problems have been examined: the feedforward control servomechanism problem (theorem 1); the perfect control problem (theorem 2); the feedforward control problem with feedback gain perturbations (theorem 4); the weak robust servomechanism problem (theorem 5); the robust servomechanism problem (theorem 6); the strong robust servomechanism problem (theorem 7); the robust control limitation theorem (theorem 8); the robust decentralized servomechanism problem (theorem 9); the sequentially reliable decentralized servomechanism problem (theorem 10); the time lag robust servomechanism problem (theorem 11); tuning regulations for unknown systems (theorem 13); and generalized error constants for multivariable systems (Section 13.4.2).

The chapter concludes with a brief description of a robust controller design method using parameter optimization methods. An example of the method is applied to design a robust controller for an unstable 4th order 2-input, 2-output system with an integrity constraint.

References

1. ASTROM, K.J., G. WITTENMARK : "On Self Tuning Regulators", Automatica, vol. 9, (1973), pp. 185-200.
2. DAVISON, E.J., S.H. WANG : "Properties and Calculation of Transmission Zeros of Linear Multivariable Time-Invariant Systems", Automatica, vol. 10, (1974), pp. 643-658.
3. DAVISON, E.J., A. GOLDENBERG : "Robust Control of a General Servomechanism Problem : the Servo Compensator", Automatica, vol. 11, (1975), pp. 461-471.
4. DAVISON, E.J., S.H. WANG : "On Pole Assignment in Linear Multivariable Systems Using Output Feedback", IEEE Trans. on Automatic Control, vol. AC-20, no. 4, (1975), pp. 516-518.
5. DAVISON, E.J. : "The Steady-State Invertibility and Feedforward Control of Linear Time-Invariant Systems", IEEE Trans. on Automatic Control, vol. AC-21, no. 4, (1976), pp. 529-534.
6. DAVISON, E.J. : "Output Detectability, Steady-State Invertibility and the General Servomechanism Problem", IEEE Conference on Decision and Control, Dec. 1976, pp. 1250-1257, (1976).
7. DAVISON, E.J. : "The Robust Control of a Servomechanism Problem for Linear Time-Invariant Multivariable Systems", IEEE Trans. on Automatic Control, vol. AC-21, no. 1, (1976), pp. 25-34.
8. DAVISON, E.J. : "The Robust Decentralized Control of a General Servomechanism Problem", IEEE Trans. on Automatic Control, vol. AC-21, no. 1, (1976), pp.14-24.
9. DAVISON, E.J. : "Multivariable Tuning Regulators: the Feedforward and Robust Control of a General Servomechanism Problem", IEEE Trans. on Automatic Control, vol. AC-21, no. 1, (1976), pp. 35-47.
10. DAVISON, E.J., S.G. CHOW : "Perfect Control in Linear Time-Invariant Multivariable Systems: The Control Inequality Principle", in Control System Design by Pole-Zero Assignment (editor: F. Fallside), Academic Press, 1977.

11. DAVISON, FERGUSON, I.J. : "Design of Controllers for the Multivariable Robust Servomechanism Problem Using Parameter Optimization Methods - Some Case Studies", 2nd IFAC Workship on Control Applications of Nonlinear Programming and Optimization, Munich, Germany, 15-17 September, 1980.

12. DAVISON, E.J., I.J. FERGUSON : "The Design of Controllers for the Multivariable Robust Servomechanism Problem Using Parameter Optimization Methods", IEEE Trans. on Automatic Control, vol. AC-20, no. 1, (1981), pp. 93-110.

13. DAVISON, E.J. : "Optimal Feedforward-Robust Controllers for the Multivariable Servomechanism Problem", Dept. of Electrical Engineering, University of Toronto, Systems Control Report No. 8105, July (1981).

14. DAVISON, E.J. : "Reliability of the Robust Servomechanism Controller for Decentralized Systems", 8th IFAC World Congress on Automatic Control, Kyoto, Japan, Aug. 24-28, (1981), to appear.

15. GOLDENBERG, A., E.J. DAVISON : "The Solvability of the Robust Parameter Feedback Servomechanism Problem", 1976 Conference on Information Sciences and Systems, April (1976), John Hopkins University, Baltimore.

16. GOLDENBERG, A., E.J. DAVISON : "The Robust Control of a Servomechanism Problem with Time Delay", 7th IFAC World Congress on Automatic Control, Helsinki, Finland, June, 1978, pp. 2095-2102.

17. SAIN, M.K., (Editor) : IEEE Trans. on Automatic Control, Special Issue on "Linear Multivariable Control Systems", vol. AC-26, no. 1, Feb. 1981.

18. SCHERZINGER, B., E.J. DAVISON : "Generalized Error Coefficients for the Multivariable Servomechanism Problem", Dept. of Electrical Engineering, University of Toronto, Systems Control Report No. 8103, June 1981, IEEE Control and Decision Conference, Dec. 1981, to appear.

19. SPANG, A., L. GERHART : GE-RPI-NSF Workshop on Control Design, Schenectady, N.Y., May 20-22, 1981.

Chapter 14

Control of distributed parameter systems

Dr. A. J. PRITCHARD

Synopsis

Much of control theory is concerned with systems which are modelled by ordinary differential equations, so-called lumped parameter systems. In this chapter it is shown how the concepts of controllability, observability, optimal control and estimation may be investigated for system models based upon partial differential equations, so-called distributed parameter systems. Such a system with a single input and a single output is used to illustrate the theory presented.

14.1 Introduction

If the state of a system at each time t cannot be described by a finite set of real numbers then we will call the system a distributed parameter system. Very important classes of distributed parameter systems are those described by partial differential equations and differential-delay equations. Since physicists and applied mathematicians have shown that many complex physical phenomena may be modelled quite accurately by distributed parameter systems it is necessary to consider how they may be controlled. Obviously it will not be possible to examine all the control theoretic topics[1,2] in this chapter but it is hoped to show that these systems pose new problems which are not insurmountable. Throughout the chapter reference will be made to the following simple example.

Example 1

Consider a metal bar in a furnace. The bar is heated by jets whose distribution $b(x)$ is fixed but whose magnitude can be varied by a control $u(t)$. For one dimensional flow, heat balance considerations yield

$$\int_{x-}^{x+} \frac{\partial\theta}{\partial t}(x,t)\,dx = \alpha\left(\frac{\partial\theta}{\partial x}(x,t)\right)_{x-}^{x+} + \int_{x+}^{x-} b(x)u(t)dx$$

where $\theta(x,t)$ is the temperature in the bar at time t, and distance x from one end. α is the thermometric conductivity. Differentiation yields

$$\frac{\partial\theta}{\partial t}(x,t) = \alpha\frac{\partial^2\theta}{\partial x^2}(x,t) + b(x)u(t)$$

Let us assume that the temperature at each end is kept at a constant value $\bar{T}$, and the initial temperature distribution in the bar is $\theta_o(x)$, so that

$$\theta(0,t) = \theta(\ell,t) = \bar{T}$$

$$\theta(x,0) = \theta_o(x)$$

where ℓ is the length of the bar. Set

$$\theta(x,t) = T(x,t) + \bar{T}$$

$$\theta_o(x) = T_o(x) + \bar{T}$$

then

$$\left.\begin{aligned} \frac{\partial T}{\partial t}(x,t) &= \alpha\frac{\partial^2 T}{\partial x^2}(x,t) + b(x)u(t) \\ T(0,t) &= T(\ell,t) = 0 \\ T(x,0) &= T_o(x) \end{aligned}\right\} \quad (14.1)$$

Finally assume that it is possible to sense the temperature at a given point and this is accomplished by taking a weighted average of the temperature at nearby points. Then the output of the sensor can be modelled by

$$h(t) = \int_0^\ell c(x)\theta(x,t)dx$$

If $$h(t) = y(t) + \bar{T}\int_0^\ell c(x)dx,$$

then

$$y(t) = \int_0^{\ell} c(x)\, T(x,t)\, dx \tag{14.2}$$

Equations (14.1), (14.2) represent a single input-single output distributed parameter system.

The results to be described are more easily formulated by abstracting the system. To do this let us ask what properties one would expect of the solution of a dynamical system on a Hilbert space, H. If it is assumed that the dynamics which govern the evolution from the initial state z_o to $z(t)$ are linear and autonomous, then for each time t, we can define a linear operator T_t, such that

$$T_o = I \tag{14.3}$$

$$z(t) = T_t\, z_o \tag{14.4}$$

Moreover, if $z(t+s)$ is the same point in H as the point reached by allowing the dynamics to evolve from $z(s)$ for a time t, we have

$$z(t+s) = T_{t+s} z_o = T_t z(s) = T_t T_s z_o$$

$$T_{t+s} = T_t\, T_s \tag{14.5}$$

If we also assume that the solution is continuous in time, then the operator T_t with properties (14.3), (14.5) is called a *strongly continuous semigroup*.

Now let us define an operator A by

$$Az = \lim_{h \to 0+} \frac{(T_h - I)z}{h}, \quad z \in D(A)$$

where $D(A)$ is the set of points for which the limit exists. Then it can be shown that (14.4) is the solution of the differential equation

$$\dot{z}(t) = Az(t), \quad z(o) = z_o \in D(A)$$

Example 1 may be cast as a control problem

$$\left.\begin{aligned} \dot{z}(t) &= Az(t) + bu(t), \quad z(o) = z_o \\ y(t) &= \langle c, z(t) \rangle \end{aligned}\right\} \qquad (14.6)$$

with solution

$$z(t) = T_t z_o + \int_0^t T_{t-\rho} bu(\rho) d\rho \qquad (14.7)$$

If the initial state is zero, then the input-output map is of the form

$$y(t) = \int_0^t \langle c, T_{t-\rho} b \rangle \, u(\rho) d\rho$$

By taking the Laplace transform we obtain

$$\bar{y}(s) = \langle c, R(s,A)\, b \rangle \; \bar{u}(s)$$

where $R(s,A)$ is the resolvent $(sI-A)^{-1}$. So the transfer function is

$$\langle c, R(s,A)\, b \rangle$$

In order to calculate this for Example 1 we need to identify the Hilbert space H as $L^2(0,\ell)$, and

$$A\dot{z} = \frac{\partial^2 z}{\partial x^2}$$

$$D(A) = (z \varepsilon H : \frac{\partial^2 z}{\partial x^2} \varepsilon H, \quad z(o) = z(\ell) = 0)$$

Then we need to solve

$$(sI-A)z = z_1$$

for $z_1 \varepsilon H$, $z \varepsilon D(A)$. This is the ordinary differential equation

$$sz(x) - \alpha\frac{d^2z}{dx^2}(x) = z_1(x)$$

$$z(0) = z(\ell) = 0$$

The solution is

$$z(x) = \frac{1}{(s\alpha)^{\frac{1}{2}}}\left\{\frac{\sinh(s/\alpha)^{\frac{1}{2}}x}{\sinh(s/\alpha)^{\frac{1}{2}}\ell}\int_0^{\ell} \sinh(s/\alpha)^{\frac{1}{2}}(\ell-\sigma)\; z_1(\sigma)d\sigma \right.$$

$$\left. - \int_0^x \sinh(s/\alpha)^{\frac{1}{2}}(x-\sigma)z_1(\sigma)d\sigma \right\}$$

So the transfer function is

$$\int_0^1 \frac{c(x)}{(s\alpha)^{\frac{1}{2}}}\left\{\frac{\sinh(s/\alpha)^{\frac{1}{2}}x}{\sinh(s/\alpha)^{\frac{1}{2}}\ell}\int_0^{\ell} \sinh(s/\alpha)^{\frac{1}{2}}(\ell-\sigma)b(\sigma)d\sigma \right.$$

$$\left. - \int_0^x \sinh(s/\alpha)^{\frac{1}{2}}(x-\sigma)b(\sigma)d\sigma\right\} dx$$

It is evident that this is not a rational function of s so that new theories for irrational transfer functions need to be developed for distributed parameter systems. An alternative approach is to show that the solution is equivalent to the solution of an infinite number of ordinary differential equations. Let us suppose that the spectrum of A consists entirely of distinct real points λ_i, $i = 1,2,\ldots.$ where

$$A\phi_i = \lambda_i \phi_i, \qquad \phi_i \in D(A) \tag{14.8}$$

If the adjoint vectors are ψ_i, so that

$$A^*\psi_i = \lambda_i \psi_i, \qquad \psi_i \in D(A^*) \tag{14.9}$$

then $\langle\phi_i, \psi_j\rangle = 0$ for $i \neq j$.

A solution of (14.6) is sought of the form

$$z(t) = \sum_{n=1}^{\infty} a_n(t)\phi_n$$

then

$$\dot{z}(t) = \sum_{n=1}^{\infty} \dot{a}_n(t)\phi_n = A\sum_{n=1}^{\infty} a_n(t)\phi_n + bu(t)$$

$$= \sum_{n=1}^{\infty} \lambda_n a_n(t)\phi_n + bu(t)$$

If we assume

$$b = \sum_{n=1}^{\infty} b_n\phi_n , \quad c = \sum_{n=1}^{\infty} c_n\psi_n$$

where $\sum_{n=1}^{\infty} b_n^2 < \infty , \quad \sum_{n=1}^{\infty} c_n^2 < \infty$

then by taking inner products with ψ_n we obtain the uncoupled system of ordinary differential equations

$$\dot{a}_n(t) = \lambda_n a_n(t) + b_n u(t) \qquad n = 1,2,\ldots.$$

$$y(t) = \sum_{n=1}^{\infty} c_n a_n(t)$$

For Example 1 we have $\psi_n(x) = \phi_n(x) = \sqrt{2/\ell}\,\sin\frac{n\pi x}{\ell}$, $\lambda_n = -n^2\pi^2\alpha/\ell^2$.

14.2 Controllability

Let us consider a question of controllability for Example 1. If the initial state is zero, we have

$$a_n(t) = \int_0^t e^{\lambda_n(t-s)}\, b_n u(s)ds$$

Suppose we wish to steer the system to a temperature distribution $T_1(x)$ at time t_1, where

$$T_1(x) = \sum_{n=1}^{\infty} t_n \phi_n(x)$$

then

$$t_n = \int_0^{t_1} e^{\lambda_n(t_1-s)} b_n u(s)ds$$

Using the Schwartz inequality

$$|t_n|^2 \leq |b_n|^2 \int_0^{t_1} e^{2\lambda_n(t_1-s)} ds \int_0^{t_1} u^2(s)ds$$

for $n = 1,2,...$

or

$$\frac{2|t_n|^2}{|b_n|^2} |\lambda_n| \leq \int_0^{t_1} u^2(s)ds \quad \text{for} \quad \lambda_n < 0$$

But if the control energy is to be finite this requires

$$|t_n| \leq \frac{K|b_n|}{|\lambda_n|^{\frac{1}{2}}} = \frac{\bar{K}|b_n|}{n}$$

for some constants K, $\bar{K}$. Thus only those states with sufficiently small Fourier coefficients can be achieved.

Definition of Exact Controllability on $(0,t_1)$:

We say that the system (14.7) or (A,b) is exactly controllable on $(0,t_1)$ if for all z_o, z_1 in the Hilbert space H there exists a control $u \in L^2(0, t_1)$ such that

$$z_1 = T_{t_1} z_o + \int_0^{t_1} T_{t_1-s} bu(s)ds$$

It can be shown that this is the case if and only if there exists a constant $\gamma > 0$, such that

$$\int_0^{t_1} |b^* T_s^* z|^2 ds \geq \gamma||z||^2 \qquad (14.10)$$

where $b^*z = \langle b,z \rangle$.

Inequality (14.10) is equivalent to

$$\langle z, \int_0^{t_1} T_s\, b\, b^* T_s z\rangle\, ds \geq \gamma ||z||^2$$

We have already seen that Example 1 cannot be exactly controllable and so this may be too strong a definition for distributed parameter systems.

Definition of Approximate Controllability on $(0, t_1)$:

System (14.7) or (A,b) is approximately controllable on $(0, t_1)$ if for all z_o and z_1 in H and any $\varepsilon > 0$, there exists a control $u \in L^2(0, t_1)$ such that

$$||z_1 - z(t_1)|| < \varepsilon$$

where $$z(t_1) = T_{t_1} z_o + \int_0^{t_1} T_{t_1 - s}\, bu(s)ds$$

It can be shown that (A,b) is approximately controllable on $(0, t_1)$ if

$$b^* T_t^* z = 0 \qquad \text{on } (0,t_1) \tag{14.11}$$

implies $$z = 0$$

For Example 1

$$b^* T_t^* z = \sum_{n=1}^{\infty} b_n \exp(-\alpha n^2 \pi^2 t/\ell^2) a_n$$

where $$z = \sum_{n=1}^{\infty} a_n \phi_n$$

So Example 1 is approximately controllable on $(0,t_1)$ if $b_n \neq 0$ for all $n = 1,2,\dots,$ where

$$b_n = \int_0^{\ell} b(x)\phi_n(x)dx$$

14.3 Observability

Corresponding to those given in Section 14.2, dual concepts of observability can be defined.

Definition of Initial Observability on $(0, t_1)$:

We say that (14.6) or (A,c) is observable on $(0, t_1)$ if

$$\langle c, T_t z_o \rangle = 0 \qquad \text{for} \quad t \in (0, t_1) \qquad (14.12)$$

implies $$z_o = 0$$

We see immediately that this is the dual concept to approximate controllability on $(0, t_1)$. In particular, Example 1 will be initially observable on $(0, t_1)$ if $c_n \neq 0$ where

$$c_n = \int_0^{\ell} c(x)\phi_n(x)dx \qquad n = 1,2,\ldots.$$

Although the initial observability implies that the map from the output to the initial state is 1-1 it may not be bounded, and this will lead to computational errors. A stronger definition is as follows:

Definition of Continuous Observability on $(0, t_1)$:

We say that (14.6) or (A,c) is continuously observable on $(0, t_1)$ if there exists a constant $\gamma > 0$ such that

$$\int_0^{t_1} |\langle c, T_s z\rangle|^2 \, ds \geq \gamma ||z||^2$$

So continuous observability is the dual concept to exact controllability and is a very strong requirement for distributed parameter systems and is certainly not satisfied by Example 1.

14.4 Optimal Control

Let us now pose the linear quadratic regulator problem, where an optimal control is sought to minimize the functional

$$J(u) = \langle z(t_1), Gz(t_1)\rangle + \int_0^{t_1} (\langle z(s), Wz(s)\rangle + Ru^2(s))ds \tag{14.13}$$

where z satisfies (14.7). We assume $G \geq 0$, $W \geq 0$, $R > 0$.

The optimal control u^* is given by

$$u^*(t) = -R^{-1} \langle b, P(t)\, z(t)\rangle$$

where $P(\cdot)$ satisfies the Riccati equation

$$\left.\begin{aligned} &\frac{d}{dt} \langle h, P(t)\, k\rangle + \langle Ah, P(t)\, k\rangle + \langle P(t)h, Ak\rangle \\ &\qquad + \langle h, Wk\rangle \\ &= R^{-1} \langle h, P(t)\, b\rangle \ \ \langle b, P(t)\, k\rangle \\ &P(t_1) = G \end{aligned}\right\} \tag{14.14}$$

So that the only difference between this and the finite dimensional case is the inner product structure of the Riccati equation. If we assume (14.8) and (14.9) hold, we may seek solutions for P in the form

$$P(t)z = \sum_{j=1}^{\infty} \sum_{k=1}^{\infty} P_{jk}(t) \ \ \langle \psi_j, z\rangle \psi_k$$

Substituting in (14.14) and setting $h = \phi_m$, $k = \phi_n$, we obtain

$$\left.\begin{aligned} &\dot{P}_{mn} + (\lambda_m + \lambda_n) P_{mn} + \langle \phi_m, W\phi_n\rangle \\ &= R^{-1} \left(\sum_{j=1}^{\infty} P_{jm} b_j\right)\left(\sum_{j=1}^{\infty} P_{nj} b_j\right) \\ &P_{mn}(t_1) = \langle \phi_m, G\phi_n\rangle \end{aligned}\right\} \tag{14.15}$$

14.5 Optimal Estimation (Kalman Filter)

If the output is corrupted by noise we may pose a state estimation problem, where the dynamics are

$$z(t) = T_t\, z_o$$

$$y(t) = \int_0^t \langle c, z(s)\rangle\, ds + v(t)$$

$v(t)$ is a Weiner process with covariance V, and z_o is Gaussian with zero mean and covariance Q_o. The state estimation problem is to find the best estimate of the state, $\hat{z}(t)$ at time t based on the observation process $y(s)$ $0 \leq s \leq t$. The solution is given by the Kalman-Bucy filter

$$\hat{z}(t) = V^{-1} \int_0^t T_{t-s} Q(s) c \, d\rho(s)$$

where $\rho(\cdot)$ is the innovations process

$$\rho(t) = y(t) - \int_0^t \langle c, \hat{z}(s) \rangle \, ds$$

and $Q(\cdot)$ satisfies the Riccati equation

$$\begin{aligned} &\frac{d}{dt} \langle Q(t) h,k \rangle - \langle Q(t)h, A^* k \rangle - \langle A^* h, Q(t)k \rangle \\ &\quad = V^{-1} \langle h, Q(t) c \rangle \langle c, Q(t) k \rangle \quad h,k \in D(A^*) \\ &Q(o) = Q_o \end{aligned} \tag{14.16}$$

If (14.8) and (14.9) hold we may seek solutions of the form

$$Q(t) = \sum_{j=1}^{\infty} \sum_{k=1}^{\infty} q_{jk}(t) \langle \phi_j, z \rangle \phi_k$$

Alternatively we may note that the transformations

$$\left.\begin{aligned} A^* &\to A \\ Q(t) &\to P(t_1 - t) \\ V &\to R \\ P_o &\to G \\ c &\to b \end{aligned}\right\} \tag{14.17}$$

send equation (14.16) into equation (14.14) and in this sense the estimation and control problems are dual to each other.

We have seen that the problems of controllability, observability, optimal control and estimation can be solved for distributed parameter systems. The added complexity of dealing with infinite dimensional systems must be balanced against the small number of physically well defined parameters which occur in the partial differential equation description and the decoupling of the modes which occurs for many systems.

References

1. BUTKOVSKIY, A.G. : Theory of Optimal Control of Distributed Parameter Systems, Elsevier, 1969.

2. CURTAIN, R.F. and A.J. PRITCHARD : Infinite Dimensional Systems Theory, Lecture Notes in Control and Information Sciences, vol. 8, Springer-Verlag, 1978.

CONTROL OF DISTRIBUTED PARAMETER SYSTEMS - Problems

P.14.1 If $b(x) = \sqrt{2/\ell}\,\sin \pi x/\ell$ in Example 1 of this chapter, find the optimal control which minimizes the performance index

$$J(u) = \int_0^{\ell} T^2(x,t_1)dx + \int_0^{t_1} u^2(t)dt$$

State a dual Kalman filtering problem.

P.14.2 Consider the controlled wave equation

$$\frac{\partial^2 w}{\partial t^2}(x,t) = \frac{\partial^2 w}{\partial x^2}(x,t) + \hat{b}(x)u(t)$$

$$y(t) = \int_0^1 \hat{c}(x)w(x,t)dx$$

with $w(0,t) = w(1,t) = 0$

Find the transfer function if the initial state is zero.
Calculate the semigroup and derive conditions for the system to be approximately controllable and initially observable.

Chapter 15

Decentralised control

Professor MADAN G. SINGH

Synopsis

In this paper we give a brief overview of some of the recent results in decentralised control. Both the stochastic decentralised control as well as the deterministic decentralised control literature are reviewed. Some of the decentralised design techniques are also outlined.

15.1 Introduction

Although decentralised controllers have been designed and used for controlling interconnected dynamical systems for over two decades, the design was based on *ad hoc* methods. For example, one usually assumed that the system comprised *weakly* interacting subsystems so that it was plausible to design the controllers independently for each subsystem. This is still the basis for most industrial control systems design. However, with the increased interest in optimisation techniques in the 1960s, there was an attempt to translate the notions of optimal centralised controller design to decentralised situations. It is only at this point that the intrinsic difficulty of decentralised control became apparent. For example, one of the best known and most useful results in the theory of control for centralised systems is the separation theorem[1]. This, broadly speaking, states that for linear dynamical systems subject to Gaussian disturbances, it is possible to design a controller which minimises the expected value of a quadric cost function by designing separately an optimal state estimator and an optimal controller. However, the state estimator is finite dimensional and the optimal controller is linear. These results fail in the case of decentralised control. To see this let us examine the information structures associated with decentralised control.

15.2 Non-classical Information Patterns

Consider a linear stochastic system described by:

$$\underline{\dot{x}}(t) = A\underline{x}(t) + \sum_{i=1}^{N} B_i\underline{u}_i(t) + \underline{\xi}(t) \quad (15.1)$$

$$\underline{y}(t) = C\underline{x}(t) + \sum_{i=1}^{N} D_i\underline{u}_i(t) + \underline{\theta}(t) \quad (15.2)$$

Here equation (15.1) is the state equation for a linear dynamical system being controlled from N control stations, equation (15.2) is the output equation. $\underline{\theta}(t)$, $\underline{\xi}(t)$ are assumed to be independent Gaussian White Noise processes. $\underline{y}(t)$ in equation (15.2) corresponds to all the measured outputs of the system. The partition of these measurements available to a given controller is defined by the *information pattern* of the problem.

For the decentralised control problem, the information pattern is defined by the set of matrices

$$H = (H_1, H_2, \ldots, H_N)$$

where $\underline{z}_i(t) = H_i(t)\underline{y}(t)$; $i = 1,2,\ldots N$

For the case of the classical information pattern, let us define

B = Block Diagonal matrix $[B_i]$, $i = 1,2,\ldots N$
D = Block Diagonal matrix $[D_i]$, $i = 1,2,\ldots N$
and H = Block Diagonal matrix $[H_i]$, $i = 1,2,\ldots N$

Then a *classical information pattern* arises when

$$H = I$$

where I is the identity matrix. This implies that all the controllers have identical information.

Now, let us consider the quadratic performance

$$J = \lim_{T\to\infty} \left\{ \frac{1}{T} \int_0^T [\underline{x}'(t)Qx(t) + \sum_{i=1}^{N} \underline{u}_i'(t)\ R_i u_i(t)dt \right\}$$

where Q is a positive semi-definite matrix and R = Block Diag $[R_i]$ is a positive definite matrix.

A *non-classical information pattern* arises if the controls are restricted to be of the form

$$\underline{u}_i(t) = \underline{\gamma}_i(\underline{z}_i^t)$$

where $\underline{z}_i^t = (\underline{z}_i(\tau) : 0 \leq \tau \leq t)$

i.e. the controls at the ith control station are chosen as some function $\underline{\gamma}_i$ of the past and present information available to it.

Witsenhausen[2] considered a seemingly simple example to show that for non-classical information patterns, the optimal control law is not linear and that the optimal estimator may be infinite dimensional. The reason for this is that due to the lack of possibility of directly transferring information between the controllers so as to reduce the overall cost, the information transfer is attempted through the dynamics of the system. Thus each controller tries to use its control action not only to exercise control but also to convey information to the other controllers. The interaction between the control and the signalling effect is necessarily non-linear!

A lot of work has been done in the last few years to seek out special cases where linear solutions are possible. Ho and Chu[3] found that the so-called "partially nested information structures" lead to decision rules which are linear in information.

15.2.1 Partially nested information structures

Consider the observations

$$y_i = H_i \xi + D_i u$$

where D_i needs to satisfy the causality conditions:

If u_i acts before u_j then $[D_i]_{\text{jth component}} = 0$

Let us consider the information structure

$$z_i = \begin{cases} z_k & k \in E_i \\ y_i & \end{cases} \qquad (15.3)$$

where $\qquad E_i = \{k | D_{ik} \neq 0\}$

The information structure equation (15.3) implies that if u_k affects the observation y_i (i.e. $D_{ik} \neq 0$) then u_i knows what u_k knows. This structure is a generalisation of perfect recall and defines a partial order of inclusion on what the various decision makers know. This is called a *partially nested information structure*. The reason why linear decision rules are applicable in this case is because *no signalling is possible*.

A lot of the work that has been done on design methods for stochastic decentralised control[4,5,6] has relied on seeking fixed structure controllers which are usually constrained to be linear. Parallel to this, a lot of work has been done on the deterministic decentralised control problem. Next we will examine the basis of this work.

15.3 Decentralised Stabilisation and Pole Placement

The deterministic decentralised control problem was attacked by Wang and Davison[7] and by Corfmat and Morse[8]. Essentially, they derived the necessary conditions for the existence of stable decentralised feedback controls.

A fundamental result in control theory, which was derived by Wonham[9] states that the poles of a closed loop system may be arbitrarily placed using state feedback provided the system is controllable. Thus controllable systems can be stabilised using a centralised controller which places the poles in the left half plane. Brasch and Pearson[10], amongst others, have extended Wonham's result to the case of output feedback. We will next see how this result generalises to the case where the feedback matrix is constrained to be block diagonal.

The system can be described by:

$$\dot{\underline{x}}(t) = A\underline{x}(t) + \sum_{i=1}^{\nu} B_i \underline{u}_i(t)$$
$$\underline{y}_i(t) = C_i\, \underline{x}(t) \qquad (15.4)$$

where ν is the number of control stations and $i = 1, \ldots N$ indexes the input and output variables of the various controllers.

The set of local feedback laws are assumed to be generated by the following feedback controllers:

$$\left.\begin{aligned} \dot{\underline{z}}_i(t) &= S_i z_i(t) + B_i y_i(t) \\ u_i(t) &= Q_i z_i(t) + K_i y_i(t) + v_i(t) \end{aligned}\right\} \qquad (15.5)$$

$$i = 1, 2, \ldots \nu$$

where $z_i(t) \in R^{n_{oi}}$ and $v_i(t) \in R^{m_i}$ is an external input.

We can rewrite equation (15.5) in a more compact form as:

$$\dot{\underline{z}}(t) = S\underline{z}(t) + R\underline{y}(t)$$
$$\underline{u}(t) = Q\underline{z}(t) + K\underline{y}(t) + \underline{v}(t) \qquad (15.6)$$

where S, Q, R, K are appropriate block diagonal matrices.

If we apply the feedback law (15.6) to the system (15.4), then the closed loop system can be represented by

$$\begin{pmatrix} \dot{\underline{x}} \\ \dot{\underline{z}} \end{pmatrix} = \left(\begin{array}{c|c} A+BKC & BQ \\ \hline RC & S \end{array}\right) \begin{pmatrix} \underline{x}(t) \\ \underline{z}(t) \end{pmatrix} + \begin{pmatrix} B \\ 0 \end{pmatrix} \underline{v}(t) \qquad (15.7)$$

where $C = (C_1^T\, C_2^T \ldots C_\nu^T)^T$; $B = [B_1, B_2, \ldots B_\nu]$

The deterministic decentralised control problem is to choose local feedback control laws (15.6) such that the overall system (15.7) is stable, i.e. all its poles are in the open left complex plane.

In order to see under what conditions one can solve this problem, we begin by defining the fixed polynomial of the triple (C,A,B).

Definition

Let $\bar{K}$ be the set of block diagonal matrices

$$\bar{K} = \{K | K = \text{block diag } [K_1, \ldots K_\nu] \}$$

where $\quad K_i \in R^{m_i \times p_i}$

Then the greatest common divisor (g.c.d.) of the set of characteristic polynomials of (A+BKC) $\forall\, K \in \bar{K}$ is called the *fixed polynomial* of the triple (C,A,B) with respect to $\bar{K}$ and it is denoted by $\psi(\lambda;C,A,B,\bar{K})$; $\psi(\lambda;C,A,B,\bar{K}) = \underset{K \in \bar{K}}{\text{g.c.d.}} \{\det(I\text{-}A\text{-}BKC)\}$

Comment

It can be shown that there exists a unique monic polynomial (i.e. a polynomial whose leading coefficient is unity) ψ such that ψ divides all the polynomials generated by $\bar{K}$. Such a polynomial is called the greatest common divisor (g.c.d.) of the set of polynomials generated by $\bar{K}$.

Definition

For a given triple (C,A,B) and the given set $\bar{K}$ of block diagonal matrices in equation (15.7), the set of *fixed modes* of (C,A,B) w.r.t. $\bar{K}$ is defined as:

$$\Lambda(C,A,B,\bar{K}) = \bigcap_{K \in \bar{K}} \sigma(A+BKC)$$

where $\sigma(A+BKC)$ denotes the set of eigenvalues of (A+BKC), i.e. of the matrix whose stability is of interest.

Alternatively, we can define $\Lambda(C,A,B,\bar{K})$ as

$$\Lambda(C,A,B,\bar{K}) = \{\lambda | \lambda \in \mathbb{C} \text{ and } \psi(\lambda;C,A,B,\bar{K}) = 0\}$$

This definition can be interpreted as follows:

Since ψ divides all polynomials, it has all the common eigenvalues shared by these polynomials.

Remark

We next see why $\Lambda(C,A,B,\bar{K})$ represents the fixed modes of the system and how these fixed modes can be computed.

Let $K \in \bar{K}$ and let $k(i,r,s)$ denote the $(r,si)^{th}$ element of K_i $(i = 1,2,\ldots \nu)$. Since $\bar{K}$ contains all block diagonal matrices, it contains the null matrix. Hence $\Lambda(C,A,B,\bar{K}) \subset \sigma(A)$. Figure 15.1 illustrates this relationship (15.8).

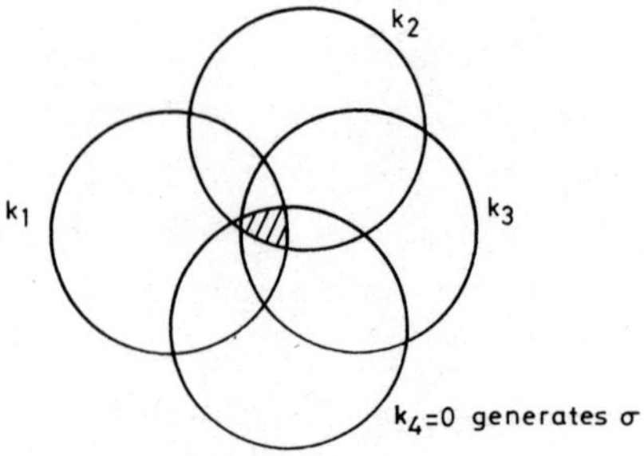

Fig. 15.1

For each $\lambda_i \in \sigma(A)$, $\det(\lambda_i I-A-BKC)$ is a polynomial in $k(i,r,s)$. If $\det(\lambda_i I-A-BKC) = 0$, then $\lambda_i \in \Lambda(C,A,B,\bar{K})$. On the other hand, if $\det(\lambda_i I-A-BKC) \neq 0$ then there exists a $\hat{K} \in K$ (at least one as shown in Figure 15.1) for which $\det(\lambda_i I-A-B\hat{K}C) \neq 0$. Hence $\lambda_i \notin \Lambda(C,A,B,\bar{K})$.

This is one way of determining the set of fixed modes of a system, i.e. we compute all the eigenvalues of A and we then apply the above test.

The reason for the name *fixed modes* is now clear, since these modes do not depend on the values assumed by $K \in \bar{K}$

but depend on the structure of $\bar{K}$ (block diagonal in our case).

Suppose we have found the eigenvalues of A that belong to Λ and these are λ_i, $i = 1,2,\ldots k$. Then we can write the g.c.d. as

$$\psi(\lambda,C,A,B,\bar{K}) = (\lambda-\lambda_1)^{\mu_1}(\lambda-\lambda_2)^{\mu_2} \ldots (\lambda-\lambda_k)^{\mu_k}$$

where μ_i is the multiplicity of λ_i amongst the eigenvalues of A.

With these preliminary definitions out of the way, we are now in a position to examine the main result on decentralised stabilisation[7].

Theorem 15.1

Consider the system (C,A,B). Let $\bar{K}$ be the set of block diagonal matrices. Then a necessary and sufficient condition for the existence of a set of local feedback control laws (15.6) such that the closed loop system (15.7) is asymptotically stable is

$$(C,A,B,K) \subset C^-$$

where C^- denotes the open left half complex plane.

This result basically says that we can only stabilise by decentralised control if the system has stable fixed modes since the latter cannot be moved by decentralised feedback.

Thus the first step in carrying out any decentralised feedback design is to check if the system has any unstable fixed modes since, in that case, it would not be possible to stabilise the system using decentralised feedback. An easy way of checking this is given by Davison[11]. First we notice that since $K_i = 0$, $i = 1,2,\ldots\nu$ is admissible, the fixed modes of A are a subset of its eigenvalues. Second, it can be shown that if K is randomly selected, then with probability one, the fixed modes of A are the common eigenvalues of A and $(A+BKC)$. Standard algorithms exist for computing the eigenvalues of fairly high order systems so the calculation of the fixed modes should not pose any difficulties.

Finally, we look at design techniques.

15.4 Design Techniques of Decentralised Control

In the case of the stochastic problem where an optimisation framework is used, due to the difficulties caused by the non-classical information patterns, all the design techniques consider fixed structure controllers[4,5,6]. Usually, this structure comprises linear feedback. Here we will concentrate on the deterministic problem. We will begin by examining the optimal approach of Geromel and Bernussou.

15.4.1 The algorithm of Geromel and Bernussou[12]

Consider the large scale system described by the state equations

$$S_i \begin{cases} \dot{\underline{x}}_i = \sum_{j=1}^{N} A_{ij}\underline{x}_j + B_i\underline{u}_i \\ \underline{x}_i(o) = \underline{x}_{io} \qquad i = 1,2,\ldots N \end{cases}$$

where $\underline{x}_i \in R^{n_i}$, $\underline{u}_i \in R^{r_i}$. This system can be rewritten in the form

$$(S) \begin{cases} \dot{\underline{x}} = A\underline{x} + B\underline{u} \\ \underline{x}(o) = \underline{x}_o \end{cases}$$

where $A = \{A_{ij},\ i = 1,\ldots N;\ \ j = 1,\ldots N\} \in R^{n\times n}$;

$B = \text{block diag } (B_1, B_2, \ldots B_N) \in R^{n\times r}$;

$$n = \sum_{i=1}^{N} n_i, \qquad r = \sum_{i=1}^{N} r_i$$

we seek a linear static decentralised feedback controller of the type

$$\underline{u}_i = - k_i x_i \; ; \quad i = 1,2,\ldots N \qquad (15.9)$$

Let $\Omega(S)$ be the set of all block diagonal gain matrices, i.e.

$$\Omega(S = \{K \in R^{r\times n} | K = \text{block diag}(k_1, k_2, \ldots k_N) \; ; \quad k_i \in R^{r_i \times n_i}\}$$

Then the problem is to find a decentralised control of the form given by equation (15.9) so as to minimise a quadratic cost function, i.e.

$$\min_{K\in\Omega(s)} \int_0^\infty (\underline{x}^T Q \underline{x} + \underline{u}^T R \underline{u}) dt$$

$$\text{s.t.} \qquad \dot{\underline{x}} = A\underline{x} + B\underline{u}$$

$$\underline{u} = - K\underline{x}$$

where it is assumed that $Q = CC^T \geq 0$ and $R > 0$.

Let $\phi(t) = \exp[A-BK]$ be the transition matrix associated with $\dot{\underline{x}} = (A-BK)\underline{x}$. Then the cost function J can be re-written as

$$J = x(o)^T \int_0^\infty \phi^T(t)(Q + K^T RK)\phi(t) dt \; x(o)$$

We can write a new cost function by considering $\underline{x}(o)$ to be a random variable which is uniformly distributed over an n-dimensional sphere of unit radius, i.e.

$$\min_{K\in\Omega(S)} \hat{J}(K) = \text{tr } P$$

$$P = \int_0^\infty \exp([A-BK]^T t)(Q+K^T BK)\exp([A-BK]^T t) dt$$

The gradient matrix $\partial\hat{J}(K)/\partial K$ is given by the system of equations:

$$\frac{\partial \hat{J}(K)}{\partial K} = 2(KK-B^TP)L$$

$$(A-BK^TP+P(A-BK)+Q+K^TRK = 0 \qquad (15.10)$$

$$(A-BK)L+L(A-BK)^T+I = 0$$

for those $K \in \Omega(S)$ such that $(A-BK)$ is a matrix with eigenvalues in the left half plane.

Let us denote by $D = \text{block diag}(D_1,D_2,\ldots D_N) \in R^{n\times r}$ a matrix which comprises those entries of $\partial J(K)/\partial K$ which correspond to the unconstrained entries of K and the other entries are zero. This is equivalent to projecting the gradient matrix on the set $\Omega(S)$, since $D \in \Omega(S)$. Then Geromel and Bernussou propose the following algorithm:

Assume that the initial matrix $K \in \Lambda(S)$ where $\Lambda(S) = \{K|K \in \Omega(S)$ and $(A-BK)$ is stable$\}$

Step 1: Using equations (15.10) determine the gradient matrix $\partial\hat{J}(K)/\partial K$ and the feasible direction matrix $D = \text{block diag}(D_1,D_2,\ldots D_N)$.

Step 2: Test for convergence by checking if $|\{d_{pq}\}_i < \varepsilon$ $\forall\, p = 1,\ldots r,\; q = 1,\ldots n_i;\; i = 1,\ldots N;$ where $\{d_{pq}\}$ is the (p,q)th entry of the matrix $D_i^{L'}$ and ε is a small pre-chosen constant. If the convergence test is satisfied, stop. Otherwise go to step 3.

Step 3: Set $K = K - aD$ where the step size $a \geq 0$ must be chosen such that $\hat{J}(K-aD) < \hat{J}(K)$. Go to step 1.

The properties of this algorithm are summarised in the following lemma:

Lemma 15.1

If $\hat{J}(K)$ is a function of class C^1 for every $K \in \Lambda(S)$, then:

1) at each iteration there exists a step size $a > 0$ such

that $\hat{J}(K-aD) < \hat{J}(K)$

2) provided that (A,C) is completely observable and $Q > 0$ then for every a which is such that $J(K-aD) < J(K) < \infty$ one has $(K-aD) \in \Lambda(S)$.

Remarks

1. The step size "a" can be determined using any conventional unidimensional search technique. The authors[12] used an adaptive step size routine. Thus if ℓ denotes the iteration index, then the step size obeys the following rule

$$a^{\ell+\ell} = \pi a^{\ell} \quad \text{if} \quad \hat{J}(K^{\ell} - a^{\ell}D^{\ell}) < \hat{J}(K)$$

and $\quad \ell(K^{\ell} - a^{\ell}D^{\ell}) > 0$

$$a^{\ell} = \nu a^{\ell} \quad \text{otherwise}$$

when $\pi > 1$, $0 < \nu < 1$ and one starts from an arbitrary initial value of a.

2. We note that it is not feasible to compute $\partial\hat{J}(K)/\partial K$ $i = 1,\ldots N$, i.e. the partial variations of $J(K)$ with respect to K_i directly. We recall from equations (15.10) that $\partial\hat{J}(K)/\partial K$ is a full matrix, although we only need the block diagonal bits of it. It would be attractive if we could calculate only the block diagonal elements. Unfortunately, this is not feasible.

3. The method needs an initial stable gain matrix K_o. This method has been applied to a number of examples[12].

Finally, we look at a different class of design techniques where the interactions between the subsystems are modelled in some way during an off-line study whilst on-line, no information is passed between the local controllers[13-18]. Due to lack of space, we will only look at one of the methods, i.e. the model following method of Hassan and Singh[13,14].

15.4.2 The model following method[13,14]

The basic problem is to minimise the cost function

$$\min J = \sum_{i=1}^{N} \frac{1}{2} \int_0^\infty \{||\underline{x}_i||^2_{Q_i} + ||\underline{u}_i||^2_{R_i}\}dt$$

$$\text{s.t.} \qquad \dot{\underline{x}}_i = A_i\underline{x}_i + B_i\underline{u}_i + C_i\underline{z}_i \tag{15.11}$$

$$\underline{z}_i = \sum_{j=1}^{N} L_{ij}x_j$$

where $\underline{x}_i$ is the n_i dimensional state vector of the ith subsystem, $\underline{u}_i$ is the m_i dimensional control vector and $\underline{z}_i$ is the q_i dimensional interconnection vector. Q_i and R_i are respectively positive semi-definite and positive definite matrices and A_i, B_i, C_i, L_{ij} are constant matrices of appropriate dimension. $||\cdot||^2_S = .T_S .$

The approach requires the introduction of a crude interaction model of the form

$$\dot{\underline{z}}_i = A_{z_i} \underline{z}_i$$

which can be used to modify the optimisation problem to

$$\min J = \frac{1}{2} \sum_{i=1}^{N} \int_0^\infty \{||\underline{y}_i||^2_{Q_i} + ||\underline{u}_i||^2_{R_i}\}dt$$

$$\text{s.t.} \qquad \dot{\underline{y}}_i = \tilde{A}_i \underline{y}_i + \tilde{B}_i \underline{u}_i \tag{15.12}$$

where

$$\underline{y}_i = \begin{bmatrix} \underline{x}_i \\ \hline \underline{z}_i \end{bmatrix} ; \quad \tilde{A}_i = \left[\begin{array}{c|c} A_i & C_i \\ \hline 0 & A_{z_i} \end{array}\right] ;$$

$$\tilde{B}_i = \begin{bmatrix} B_i \\ \hline 0 \end{bmatrix} ; \quad \tilde{Q}_i = \left[\begin{array}{c|c} Q_i & 0 \\ \hline 0 & 0 \end{array}\right]$$

Then the control is given by

$$\underline{u}_i = R_i^{-1}\tilde{B}_i^T P_i \underline{y}_i \tag{15.13}$$

It should be recognised that there is no simple and infallible way of choosing the crude model by picking out certain entries from the A matrix. Another way of choosing A_{z_i} may be to concentrate on the eigenvalues. However, in the multivariable situation, the relationship between the modes and the states is not a clear one.

If we recall the basic notion of the model follower for centralised control[20] we start with a certain model and then we minimise the errors between the plant output and the model output. Thus, the choice of the model is largely governed by the response that we desire from the plant.

A simple way of specifying this response is in terms of assignment of the poles of the model in the left half plane. Thus, we could choose the matrices A_{z_i} to be diagonal with negative elements.

Let us rewrite the control of $\underline{u}_i^*$ given in equation (15.13) as

$$\underline{u}_i^* = - G_{i1}\,\underline{x}_i - G_{i2}\,\underline{z}_i \tag{15.14}$$

Then we construct a subsystem model which is forced by the interconnections, i.e.

$$\dot{\hat{\underline{x}}}_i = A_i\hat{\underline{x}}_i + B_i\underline{u}_i + C_i\hat{\underline{z}}_i \tag{15.15}$$

where $\hat{\underline{x}}_i$ is the state vector of the subsystem model and $\hat{\underline{z}}_i$ is a "desired" interaction input which is consistent with any *a priori* knowledge that we have about the system. Also we substitute for $\underline{u}_i$ in equations (15.11) and (15.15) after having replaced $\underline{z}_i$ by $\hat{\underline{z}}_i$ in equations (15.14)

$$\dot{\underline{x}}_i' = (A_i - B_iG_{i1})\underline{x}_i' + C_i\underline{z}_i - B_iG_{i2}\hat{\underline{z}}_i$$

or

$$\dot{\underline{x}}_i' = \hat{A}_i\underline{x}_i' + C_i\underline{z}_i - B_iG_{i2}\hat{\underline{z}}_i \quad (15.16)$$

where $\hat{A}_i = A_i - B_iG_{i1}$

$\underline{x}_i'$ is the suboptimal state of subsystem i resulting from using $\hat{\underline{z}}_i$ instead of $\underline{z}_i$ and

$$\dot{\hat{\underline{x}}}_i' = \hat{A}_i\hat{\underline{x}}_i' + C_i\underline{z}_i - B_iG_{i2}\hat{\underline{z}}_i \quad (15.17)$$

Then in order to minimise the errors between subsystem i and subsystem model, let

$$\tilde{\underline{x}}_i = \underline{x}_i' - \hat{\underline{x}}_i'$$
$$\tilde{\underline{z}}_i = \underline{z}_i - \hat{\underline{z}}_i \quad (15.18)$$

Subtracting equation (15.17) from equation (15.16) we obtain

$$\dot{\tilde{\underline{x}}}_i = \hat{A}_i\tilde{\underline{x}}_i + C_i\tilde{\underline{z}}_i \quad (15.19)$$

Write $$\min J_i = \frac{1}{2}\int_0^\infty (||\tilde{\underline{x}}_i||^2_{H_i} + ||\tilde{\underline{z}}_i||^2_{S_i})dt$$

subject to the constraint (15.19), where H_i and S_i are positive semi-definite and positive definite matrices respectively.

The solution of this problem is given by

$$\tilde{\underline{z}}_i^* = - S_i^{-1}C_i^TK_i\tilde{\underline{x}}_i$$

where K_i is a solution of an appropriate Riccati equation.

Using equation (15.18), we obtain

$$\underline{z}_i^* = \hat{\underline{z}}_i - S_i^{-1}C_i^TK_i\tilde{\underline{x}}_i$$

Thus, the dynamical equation for the subsystem will be

$$\dot{\underline{x}}_i = \hat{A}_i \underline{x}_i + C_i \underline{z}_i - B_i G_{i2}[\hat{\underline{z}}_i - S_i^{-1} C_i^T K_i (\underline{x}_i - \hat{\underline{x}}_i)]$$

$$= (\hat{A}_i + B_i G_{i2} S_i^{-1} C_i^T K_i)\underline{x}_i + C_i \underline{z}_i - B_i G_{i2}(\hat{\underline{z}}_i + S_i^{-1} C_i^T K_i \hat{\underline{x}}_i)$$

$$\dot{\hat{\underline{x}}}_i = \hat{A}_i \hat{\underline{x}}_i + (C_i - B_i G_{i2})\hat{\underline{z}}_i$$

$$\dot{\hat{\underline{z}}}_i = A_{z_i} \hat{\underline{z}}_i$$

and we could represent these equations by Figure 15.2 .

Now further analysis can proceed with the block diagram in Figure 15.2 .

First, no matter how A_{z_i} is chosen, the errors will be corrected to a large extent by the feedback channel. Of course, the accuracy will depend partly on the numbers in A_{z_i} . Here, $\underline{z}_i$ produced by the minor loop I acts as a reference input. It is independent of any controls that could be applied. The same notion is given by (15.12) where the left side of A_{z_i} is zero, and the lower part of B is also zero. This means that no controls or other states can effect $\underline{z}_i$. For stabilising the minor loop I, we should choose real negative diagonal elements in A_{z_i} . If one does not choose a diagonal matrix, at least one must ensure that all the eigenvalues of A_{z_i} have negative real parts, and they match the required responses.

This method has also been tested on a number of examples including that of the energy cycle of a ship as well as for river pollution control[13,14].

15.5 Concluding Remarks

In this chapter we have given a brief overview of the recent research that has been done in the area of Decentralised Control. Essentially, the work can be split into three broad categories, i.e. stochastic decentralised control, deterministic stabilisation and pole placement and design techniques. We have outlined some of the more important approaches in each of these areas. A more detailed treatment of the whole field is given in the recent book[19].

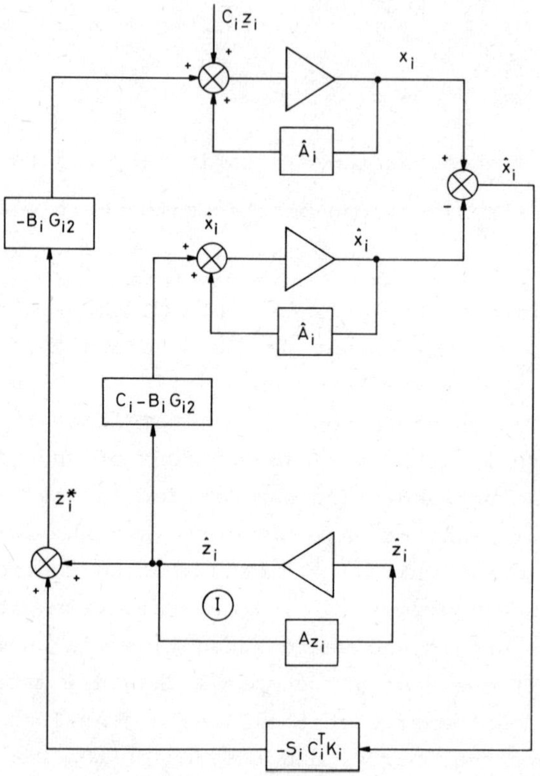

Fig. 15.2

References

1. JOSEPH, P. and J. TOU : "On linear control theory", Trans. AIEE (Appl. and Industry), vol. 80, (1961), 193-196.

2. WITSENHAUSEN, S. : "A counter-example in stochastic optimal control", SIAM J. of Control, vol. 6, (1968) 1.

3. HO, Y.C. and K.C. CHU : "Team decision theory and information structures in optimal control problems Parts I and II", IEEE Trans. AC-17, (1972), 15-28.

4. SANDELL, N. and M. ATHANS : "Solution of some nonclassical LQG stochastic decision problems", IEEE Trans. AC-18, (1974), 108-116.

5. AOKI, M. : "On decentralised linear stochastic control problems with quadratic cost", IEEE Trans. AC18, (1973), 243-250.

6. CHONG, C. and M. ATHANS : "On the stochastic control of linear systems with different information sets", IEEE Trans AC16, (1971), 423-430.

7. WANG, S. and E.J. DAVISON : "On the stabilisation of decentralised control systems", IEEE Trans AC18, (1973), 473-478.

8. CORFMAT, J. and A. MORSE : "Control of linear systems through specified input channels", SIAM J. Control, (1976), 14.

9. WONHAM, N. : "On pole assignment in multi-input controllable linear systems", IEEE Trans AC12, (1967), 660-665.

10. BRASCH, F. and PEARSON, J. : "Pole placement using dynamic compensators", IEEE Trans AC5, (1970), 34-43.

11. DAVISON, E.J. : "Decentralised stabilisation and regulation" in "Directions in large scale systems", Y.C. Ho and S. Mitter, Eds., New York Plennum, 1976.

12. GEROMEL, J. and J. BERNUSSOU : "An algorithm for optimal decentralised regulation of linear-quadratic interconnected systems", Automatica 15, (1979), 489-491.

13. HASSAN, M.F. and M.G. Singh : "A decentralised controller with on-line interaction trajectory improvement", Proc. IEE Series D, Vol. 127, 3, (1980) 142-148.

14. CHEN YULIU and M.G. SINGH : "Certain practical considerations in the model following method of decentralised control", Proc. IEE, D, vol. 128, (1981).

15. HASSAN, M.F. and M.G. SINGH : "A hierarchical structure for computing near optimal decentralised control", IEEE Trans. SMC 8, 7, (1978), 575-579.

16. HASSAN, M.F. and M.G. SINGH : "Robust decentralised controller for linear interconnected dynamical systems", Proc. IEE, Vol. 125, 5, (1978), 429-432.

17. HASSAN, M.F., M.G. SINGH and A. TITLI : "Near optimal decentralised control with a fixed degree of stability", Automatica 15, (1979), 483-488.

18. ARMENTANO, V.A. and M.G. SINGH : "A new approach to the decentralised controller initialisation problem", Proc. 8th IFAC World Congress, Kyoto, 1981.

19. SINGH, M.G. : "Decentralised Control", North Holland, Amsterdam, 1981.

20. MARKLAND, C. : "Optimal model following control system synthesis technique", Proc. IEE, vol. 117,(1980),3.

Chapter 1 : Solutions

S.1.1 $H = [b_1, Ab_1, b_2, Ab_2] = \begin{bmatrix} 1 & -2 & 0 & 2 \\ 0 & -2 & 1 & 0 \\ 0 & -2 & 1 & -1 \\ 1 & -1 & 0 & -2 \end{bmatrix}$

$$\therefore \quad H^{-1} = \begin{bmatrix} -1 & -2 & 2 & 2 \\ -1 & 0 & 0 & 1 \\ -2 & 1 & 0 & 2 \\ 0 & 1 & -1 & 0 \end{bmatrix}$$

$$\therefore \quad \begin{bmatrix} g_1^T \\ g_2^T \end{bmatrix} = \begin{bmatrix} -1 & 0 & 0 & 1 \\ 0 & 1 & -1 & 0 \end{bmatrix}$$

$$\therefore \quad Q^T = \begin{bmatrix} g_1^T \\ g_1^T A \\ g_2^T \\ g_2^T A \end{bmatrix} = \begin{bmatrix} -1 & 0 & 0 & 1 \\ 2 & 0 & 0 & -1 \\ 0 & 1 & -1 & 0 \\ 0 & 0 & 1 & 0 \end{bmatrix}$$

$$\therefore \quad (Q^T)^{-1} = \begin{bmatrix} 1 & 1 & 0 & 0 \\ 0 & 0 & 1 & 1 \\ 0 & 0 & 0 & 1 \\ 2 & 1 & 0 & 0 \end{bmatrix}$$

$$\therefore \quad Q^T A (Q^T)^{-1} = \begin{bmatrix} 0 & 1 & 0 & 0 \\ -2 & -3 & 1 & 2 \\ 0 & 0 & 0 & 1 \\ -3 & -2 & 0 & -1 \end{bmatrix}, \quad Q^T B = \begin{bmatrix} 0 & 0 \\ 1 & 0 \\ 0 & 0 \\ 0 & 1 \end{bmatrix},$$

$$C(Q^T)^{-1} = \begin{bmatrix} 0 & 0 & -1 & -2 \\ 3 & 2 & 0 & 0 \end{bmatrix}$$

S.1.2 $R_3 = I, \quad a_3 = -\operatorname{tr}(A) = 4,$

$$R_2 = A + 4I = \begin{bmatrix} 2 & 1 & 1 & 0 \\ -1 & 4 & 0 & -1 \\ -1 & 0 & 3 & -1 \\ 0 & 1 & 1 & 3 \end{bmatrix}, \quad a_2 = -\tfrac{1}{2}\operatorname{tr}(AR_2) = 9,$$

$$R_1 = AR_2+9I = \begin{bmatrix} 3 & 2 & 1 & -2 \\ -2 & 7 & -2 & -3 \\ -1 & -2 & 4 & -2 \\ -2 & 3 & 2 & 4 \end{bmatrix}, \quad a_1 = -\tfrac{1}{3}\operatorname{tr}(AR_1) = 10,$$

$$R_0 = AR_1+10I = \begin{bmatrix} 1 & 1 & 0 & -1 \\ -1 & 5 & -3 & -2 \\ 0 & -3 & 3 & 0 \\ -1 & 2 & 0 & 1 \end{bmatrix}, \quad a_0 = -\tfrac{1}{4}\operatorname{tr}(AR_0) = 3$$

$$\therefore \quad G(s) = \frac{1}{s^4+4s^3+9s^2+10s+3} \begin{bmatrix} 4s^2+8s+3 & -(2s^3+7s^2+7s+2) \\ 2s^3+5s^2+3s & 4s^2+8s+3 \end{bmatrix}$$

Chapter 3: Solutions

S.3.1 For the nominal controller parameters the input and output return-ratio matrices are both equal to G(s) which can be diagonalized at all frequencies using a constant matrix W as follows:

$$WG(s)W^{-1} = \begin{pmatrix} \frac{1}{s+1} & 0 \\ 0 & \frac{2}{s+2} \end{pmatrix}$$

where $$W = \begin{pmatrix} 7 & 8 \\ 6 & 7 \end{pmatrix}$$

and $$W^{-1} = \begin{pmatrix} 7 & -8 \\ -6 & 7 \end{pmatrix}$$

The characteristic gain loci are therefore given explicitly as

$$\frac{1}{1+j\omega} \quad \text{and} \quad \frac{2}{2+j\omega}$$

The columns of W are a candidate set of right-hand eigenvectors for G(s), and the rows of W^{-1} are a candidate set of left-hand eigenvectors. By normalizing the appropriate vectors to unity it follows that

$$\underline{x}_1 = \frac{1}{\sqrt{85}}\begin{pmatrix} 7 \\ 6 \end{pmatrix}, \quad \underline{x}_2 = \frac{1}{\sqrt{113}}\begin{pmatrix} 8 \\ 7 \end{pmatrix}$$

$$\underline{y}_1 = \frac{1}{\sqrt{113}}\begin{pmatrix} 7 \\ -8 \end{pmatrix}, \quad \underline{y}_2 = \frac{1}{\sqrt{85}}\begin{pmatrix} 6 \\ -7 \end{pmatrix}$$

and therefore that the sensitivity indices are given (at all frequencies) by

$$|s_1| = |\underline{y}_1^T \underline{x}_1| = \frac{1}{\sqrt{85}\ \sqrt{113}} \simeq 0.0102$$

$$|s_2| + |\underline{y}_2^T \underline{x}_2| = \frac{1}{\sqrt{85}\ \sqrt{113}} \simeq 0.0102$$

The very small indices indicate that the characteristic gain loci are very sensitive to small perturbations either before or after the plant, and indeed it can be shown that there are small perturbations of k_1 and k_2 for which the closed-loop system becomes unstable. Note that for $k_1 = k_2 = k$ the system has infinite gain margins with respect to k.

S.3.2 The small gain theorem tells us that the feedback design remains stable if the perturbation $\Delta G(s)$ is stable and if the maximum principal gain of $\Delta G(j\omega)$ is less than the reciprocal of the maximum principal gain of the output relative stability matrix $R_o(j\omega)$.

At low frequencies this implies that the maximum principal gain of $\Delta G(j\omega)$ is less than 1 for stability to be maintained. This limit is reduced to about 0.9 as ω increases in the medium frequency range; but at higher frequencies, as the gains roll off, clearly much larger perturbations can be accommodated.

The conservativeness of the small gain theorem at low frequencies can be reduced by considering the small phase theorem. For example, up to about $1 rs^{-1}$ the condition number of $R_o(j\omega)$ is approximately 1, $\psi_m(\omega)$ is approximately 0, and so stability is maintained (in this frequency range) for large perturbation condition numbers providing the minimum principal phase (maximum phase lag) of the perturbation is greater than say -140°. Note that insofar as stability is concerned we restrict only the condition number of the perturbation and therefore in theory the perturbation can have infinitely large gains.

A satisfactory characterization of robustness is obtained by combining the small gain and phase theorems as in Corollary 2, where, for example, one might take ω_b to be $10rs^{-1}$.

Chapter 5: Solutions

S.5.1 We have

$$|T_H| = |T_G||\hat{H}|/|\hat{G}|$$

$$= (s-1)(s+1)^2 \begin{vmatrix} (s+1)+1 & 0 \\ 0 & (s-1)+2 \end{vmatrix} \div \begin{vmatrix} s+1 & 0 \\ 0 & s-1 \end{vmatrix}$$

$$= \frac{(s-1)(s+1)^3(s+2)}{(s-1)(s+1)} = (s+1)^2(s+2)$$

so the closed-loop system is asymptotically stable.

G can arise from

$$P_o(s) = \left[\begin{array}{cc|cc} s+1 & 0 & 1 & 0 \\ 0 & s-1 & 0 & 1 \\ \hline -1 & 0 & 0 & 0 \\ 0 & -1 & 0 & 0 \end{array}\right]$$

which has least order $n = 2$ and has $|T_o| = (s-1)(s+1)$. But $G(s)$ actually arises from a 3rd order system with $|T_G| = (s-1)(s+1)^2$. Hence $s = -1$ a decoupling zero of the system giving rise to G.

The INA degenerates to the two inverse Nyquist plots, of

$$\hat{g}_{11}(s) = 1+s, \qquad \hat{g}_{22}(s) = -1+s$$

The first gives $\hat{N}_{g1} = 0$, $\hat{N}_{h1} = 0$. The second gives $\hat{N}_{g2} = 1$, $\hat{N}_{h2} = 0$. Also $p_o = 1$, so

$$\Sigma\hat{N}_{gi} - \Sigma\hat{N}_{hi} = 1 = p_o$$

whence the system is asymptotically stable.

S.5.2

(i) For a conventional feedback system with unity feedback,

$$\underline{y}(s) = H(s)\,\underline{r}(s)$$

where $$H(s) = [I+G(s)K(s)]^{-1}\,G(s)K(s)$$

i.e. $H(s)$ is a complicated function of $G(s)$ and $K(s)$. However, if the system is functionally controllable; i.e. $G^{-1}(s)$ exists; then

$$H^{-1}(s) = K^{-1}(s)G^{-1}(s) + I$$

which is simpler to deal with. Thus, Rosenbrock's INA design method is carried out using the inverse system representations.

(ii)

$$G(s) = \begin{pmatrix} \frac{1}{s+1} & 0 \\ 0 & \frac{1}{s+2} \end{pmatrix} \begin{pmatrix} 1 & -1 \\ 1 & 1 \end{pmatrix}$$

Therefore,

$$G^{-1}(s) = \frac{1}{2}\begin{pmatrix} 1 & 1 \\ -1 & 1 \end{pmatrix}\begin{pmatrix} s+1 & 0 \\ 0 & s+2 \end{pmatrix}$$

$$= \frac{1}{2}\begin{pmatrix} (s+1) & (s+2) \\ -(s+1) & (s+2) \end{pmatrix}$$

with corresponding INA for $s = jw$ as shown.

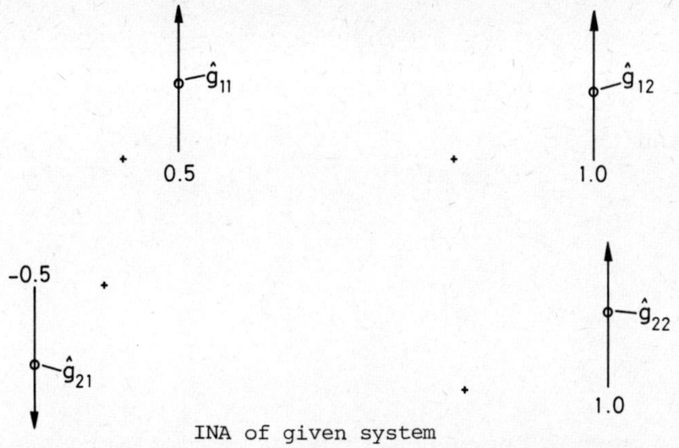

INA of given system

An examination of this inverse Nyquist array reveals that for row diagonal dominance - row 1 is not dominant
row 2 is dominant

and for column diagonal dominance - col 1 is not dominant
col 2 is not dominant

although in the latter case, we are on the margin between dominance and non-dominance.

In either case, dominance can be determined graphically by superimposing circles of radius

$$d_i(s) \;=\; |\hat{g}_{ij}(s)| \qquad i \neq j, \;\; i = 1,2$$

on each diagonal element for row dominance, and circles of radius,

$$d'_j(s) \;=\; |\hat{g}_{ij}(s)| \qquad i \neq j, \;\; j = 1,2$$

on each diagonal element for column dominance.

(iii) For this particular system, it is possible to determine a wholly real compensator $\hat{K}$ which will diagonalise the resulting system $\hat{Q}(s) = \hat{K}\hat{G}(s)$.

The required compensator can be readily determined by inspecting $\hat{G}(s)$ as

$$\hat{K} = \begin{pmatrix} 1 & -1 \\ 1 & 1 \end{pmatrix}$$

which results in

$$\hat{Q}(s) = \begin{pmatrix} s+1 & 0 \\ 0 & s+2 \end{pmatrix}$$

Thus, $K = \hat{K}^{-1} = \frac{1}{2}\begin{pmatrix} 1 & 1 \\ -1 & 1 \end{pmatrix}$

(iv) If we now introduce integral action into the two control loops for the system

$$Q(s) = \begin{pmatrix} \frac{1}{s+1} & 0 \\ 0 & \frac{1}{s+2} \end{pmatrix}$$

we can sketch the behaviour of the two decoupled responses for varying proportional gains as shown below

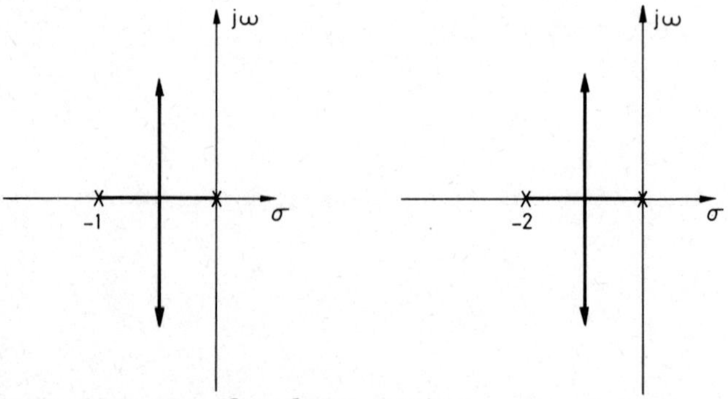

Root-locus diagrams for loop tuning

Thus, the final loop turning can be set by the appropriate choice of k_1 and k_2 to give the desired time responses for the two outputs y_1 and y_2.

Chapter 7: Solutions

S.7.1 $$H = \lambda_1 x_2 + \lambda_2(-x + u)$$

Canonical equations

$$\dot{\lambda}_1 = -H_{x_1} = \lambda_2$$

$$\dot{\lambda}_2 = -H_{x_2} = -\lambda_1$$

Then $$\lambda_2 = A\sin(t-\alpha), \quad A > 0$$

Pontryagin's principle yields $u = -\mathrm{sgn}\,\lambda_2 = -\mathrm{sgn}(\sin(t-\alpha))$

Thus, control function $u(t)$ alternates between +1 and -1.
For $u = 1$,

$$\dot{x}_1 = x_2, \quad \dot{x}_2 = -x_1 + 1$$

i.e. $$\frac{dx_1}{dx_2} = \frac{x_2}{-x_1+1} \quad \text{whence} \quad (x_1-1)^2 + x_2^2 = R^2$$

i.e. Circles with centre (1, 0), radius R.

Also, $$x_1 - 1 = -R\cos(t+\gamma) \quad \text{and} \quad x_2 = R\sin(t+\gamma)$$

$$\therefore \quad \dot{x}_1^2 + \dot{x}_2^2 = R^2$$

Thus, phase point moves clockwise with uniform velocity R.
One revolution therefore takes a time of 2π.
Similarly for $u = -1$

$$\dot{x}_1 = x_2, \quad \dot{x}_2 = -x_1 - 1$$

Phase trajectories are $(x_1+1)^2 + x_2^2 = R^2$: Circles centre (-1, 0)

If A is the point (-1,2) then choose $u = -1$ and travel on circle, centre (-1, 0), radius 2 until intersection of circle, centre (1,0), radius 1, at point B in fourth quadrant. Then choose $u = 1$ from B to origin. Time from A to B clearly less than π. Time from B to origin less than $\pi/2$. Total time therefore less than $3\pi/2$.

S.7.2 For scalar control : $(-1)^q \frac{\partial}{\partial u} H_u^{(2q)} \geq 0$ (order q)

For vector control : $(-1)^q \frac{\partial}{\partial u} H_u^{(2q)}$; $\frac{\partial}{\partial u} H_u^{(p)} = 0$

(p odd).

$$H = \lambda_1 \cos x_3 + \lambda_2 \sin x_3 + \lambda_3 u$$

$$\dot{\lambda}_1 = -H_{x_1} = 0, \quad \dot{\lambda}_2 = -H_{x_2} = 0$$

$$\dot{\lambda}_3 = -H_{x_3} = \lambda_1 \sin x_3 - \lambda_2 \cos x_3 ,$$

$$\lambda_3(t_f) = 0$$

If $\lambda_3 > 0$ then $u = -1$

$\lambda_3 < 0$ $u = +1$

$\lambda_3 \equiv 0$ $\tan x_3 = \lambda_2/\lambda_1$ (singular arc)

Suppose $u = +1$. Then $x_3 = t$, $x_2 = 1-\cos t$, $x_1 = \sin t$.

Suppose $\lambda_3 \equiv 0$. Then $\dot{\lambda}_3 \equiv 0$ so $\tan x_3 \equiv \lambda_2/\lambda_1$ (*)

$\lambda_1 = a$, $\lambda_2 = b$, a and b constants

Differentiating (*) : $u \sec^2 x_3 \equiv 0$ so $\underline{u \equiv 0}$

Then $x_3 = A$, $x_2 = t \sin A + B$, $x_1 = t \cos A + C$.

With a junction at $t = \frac{\pi}{4}$, we have $A = \frac{\pi}{4}$, $B = 1 - \frac{1}{\sqrt{2}}(1+\frac{\pi}{4})$

$C = \frac{1}{\sqrt{2}}(1 - \frac{\pi}{4})$. Also, $\tan \frac{\pi}{4} = 1 = \frac{\lambda_2}{\lambda_1} = \frac{b}{a}$

$\therefore$ $a = b$.

$$H(t_f) = -\frac{\partial G}{\partial t_f} , \quad G = t_f \quad \therefore \quad a + b = -\sqrt{2}$$

Then $a = b = -\frac{1}{\sqrt{2}}$.

On singular subarc

$$x_3 = \frac{\pi}{4}, \quad x_2 = 1 + \frac{1}{\sqrt{2}}(t - 1 - \frac{\pi}{4})$$

$$x_1 = \frac{1}{\sqrt{2}}(t + 1 - \frac{\pi}{4})$$

$x_1(t_f) = \sqrt{2}$ and $x_2(t_f) = 1$ satisfied when

$$t = 1 + \frac{\pi}{4}$$

All constraints satisfied with $u = 1,\ t \leq \frac{\pi}{4}$ and $u = C.\ \frac{\pi}{4} < t \leq 1 + \frac{\pi}{4}$.

$$H_u = \lambda_3, \quad \dot{H}_u = \dot{\lambda}_3 = \lambda_1 \sin x_3 - \lambda_2 \cos x_3$$

$$\ddot{H}_u = (\lambda_1 \cos x_3 + \lambda_2 \sin x_3)u \ ;$$

$$\frac{\partial}{\partial u}\ddot{H}_u = -\frac{1}{\sqrt{2}}\left(\frac{1}{\sqrt{2}} + \frac{1}{\sqrt{2}}\right) = -1 < 0$$

Chapter 10: Solutions

S.10.1 We want to shift the two complex eigenvalues $\{\lambda_i\} = \{+j,-j\}$ to $\{\gamma_i\} = \{-2,-2\}$ and to leave $\lambda_3 = -2$ unaltered; here, $q = 2$. If we arbitrarily select the elements of $\underline{f}$ as

$$\underline{f} = \begin{pmatrix} 1 \\ 1 \end{pmatrix}$$

then

$$\underline{b}_f = B\underline{f} = \begin{pmatrix} 0 \\ 1 \\ 1 \end{pmatrix}$$

and

$$\Phi_c = (\underline{b}_f,\ A\underline{b}_f,\ A^2\underline{b}_f)$$

$$= \begin{pmatrix} 0 & 3 & -2 \\ 1 & 0 & 1 \\ 1 & 2 & -2 \end{pmatrix}$$

which has rank 3 = n. Therefore, this choice of $\underline{f}$ results in $[A, \underline{b}_f]$ being completely controllable.

Before using the relationship given in equation (10.8) to determine the weighting factors δ_j, first determine the scalars p_j as given by equation (10.9);

$$p_1 = \langle \underline{v}_1, \underline{b}_f \rangle = -4 - j3$$

$$p_2 = \langle \underline{v}_2, \underline{b}_f \rangle = -4 + j3$$

Now, using (10.8), we obtain

$$\delta_1 = \frac{(\lambda_1-\gamma_1)(\lambda_1-\gamma_2)}{p_1(\lambda_1-\lambda_2)} = \frac{-.5(3 + j4)}{(3 - j4)}$$

$$= -0.14 + j0.48$$

Similarly, $\delta_2 = -0.14 - j0.48$

Thus, the required measurement vector $\underline{m}^t$ is given by

$$\underline{m}^t = \delta_1 \underline{v}_1^t + \delta_2 \underline{v}_2^t = (-1.4\ 0.4\ 3.6)$$

and the required state-feedback matrix is then

$$K = \underline{f}\ \underline{m}^t = \begin{pmatrix} -1.4 & 0.4 & 3.6 \\ -1.4 & 0.4 & 3.6 \end{pmatrix}$$

This results in

$$A-BK = \begin{pmatrix} -2 & 1 & 2 \\ 0.4 & -2.4 & -1.6 \\ -0.6 & -0.4 & -1.6 \end{pmatrix}$$

which has the desired eigenvalues $\{\gamma_i\} = \{-2, -2, -2\}$.

S.10.2 Using Young's algorithm on the system of Problem 1, we obtain

$$\Delta_o(s) = |sI-A| = s^3 + 2s^2 + s + 2$$

and since the desired closed-loop system eigenvalues are

$$\{\gamma_i\} = \{-2, -2, -2\}$$

we have $$\Delta_d(s) = s^3 + 6s^2 + 12s + 8$$

Therefore, the difference polynomial is

$$\delta(s) = 4s^2 + 11s + 6$$

Now, for $\underline{f} = (1\ \ 1)^t$ we know that $[A, \underline{b}_f]$ is completely controllable with Φ_c as in Problem 1. The coefficient matrix X is

$$X = \begin{pmatrix} 1 & 0 & 0 \\ 2 & 1 & 0 \\ 1 & 2 & 1 \end{pmatrix}$$

and using the relationship given in equation (10.13), we determine $\underline{m}^t$ as

$$\underline{m}^t = [\Phi_c^{t-1}]\, X^{-1}\, \underline{\delta} = (-1.4 \quad 0.4 \quad 3.6)$$

which is the same result as that achieved using the spectral approach. Thus, the resulting closed-loop system A-matrix $(A-B\underline{f}\ \underline{m}^t)$ has eigenvalues $\{\gamma_i\} = \{-2, -2, -2\}$.

S.10.3 For $(\tilde{A}-\tilde{B}\tilde{K})$ to have eigenvalues $\{\gamma_i\} = \{-4, -4, -4\}$ we proceed as follows. The matrix $\tilde{B}\tilde{K}$ has the parametric form

$$\tilde{B}\tilde{K} = \begin{pmatrix} 0 & 0 & 0 \\ k_{11} & k_{12} & k_{13} \\ k_{21} & k_{22} & k_{23} \end{pmatrix}$$

So, by choosing $k_{13} = 10/9$, $k_{21} = k_{22} = 0$, the matrix $(\tilde{A}-\tilde{B}\tilde{K})$ becomes block lower triangular, and its eigenvalues are determined by the roots of

$$\Delta_o(s) = \begin{vmatrix} 0 & -1 \\ -80/9+k_{11} & s-26/3+k_{12} \end{vmatrix} \begin{vmatrix} s - \frac{1}{3} + k_{23} \end{vmatrix}$$

Therefore, if we express $\Delta_d(s)$ as

$$\Delta_d(s) = (s^2 + 8s + 16)(s+4)$$

then the desired result can be achieved with

$$k_{11} = 224/9, \quad k_{12} = 50/3, \quad k_{23} = 13/3$$

i.e.

$$\tilde{K} = \begin{pmatrix} 224/9 & 50/3 & 10/9 \\ 0 & 0 & 13/3 \end{pmatrix}$$

which has full rank 2 = m.

The required state-feedback matrix K in the original basis is then

$$K = Q^{-1}\, \tilde{K}\, T$$

$$= \begin{pmatrix} 1 & 4/3 \\ 0 & 1 \end{pmatrix} \tilde{K} \begin{pmatrix} 0 & 0 & 1/3 \\ 1 & 4/3 & 5/3 \\ 0 & 1 & -2/3 \end{pmatrix}$$

$$= \begin{pmatrix} 50/3 & 158/9 & 1058/27 \\ 0 & 13/3 & -26/9 \end{pmatrix}$$

which has rank 2, as expected.

S.10.4 Since $m+\ell-1 < n$, a compensator of degree $r = 1$ is required for arbitrary pole assignment. However, for any desired pole configuration, the minimum degree of compensator required is given by the smallest r such that

$$\rho\{X_r\} = \rho\{X_r, \underline{\delta}_r\}$$

For $\{\gamma_i\} = \{-1, -2, -4\}$,

$$\Delta_d(s) = s^3 + 7s^2 + 14s + 8$$

and since $\Delta_o(s) = s^3 + 6s^2 + 7s + 2$

we have (for $r = 0$)

$$\underline{\delta}_o = (1 \quad 7 \quad 6)^t$$

Here,

$$X_o = \begin{pmatrix} 1 & 0 \\ 6 & 1 \\ 4 & 2 \end{pmatrix}$$

which has rank = 2, and since $\{X_o, \underline{\delta}_o\}$ also has rank = 2 we have that $r = 0$ is both necessary and sufficient for the desired closed-loop pole set $\{\gamma_i\} = \{-1, -2, -4\}$.

Now, a g_1-inverse of X_o is

$$X_o^{g_1} = \begin{pmatrix} 1 & 0 & 0 \\ -6 & 1 & 0 \end{pmatrix}$$

therefore, $\underline{p}_o = X_o^{g_1}\,\underline{\delta}_o = \begin{pmatrix} 1 \\ 1 \end{pmatrix}$

Thus, the required feedback is $F = [1 \quad 1]$.

Now, for $\{\gamma_i\} = \{-1, -2, -3\}$

$$\Delta_d(s) = s^3 + 6s^2 + 11s + 6$$

which yields $\underline{\delta}_o = (0 \quad 4 \quad 4)^t$

However, for this closed-loop pole specification $\rho\{X\} \neq \rho\{X_o, \underline{\delta}_o\}$ and we must increase r from 0 to 1. Since the resulting closed-loop system has 4 poles, we consider $\{\gamma_i\} = \{-1, -2, -3, -4\}$ with

$$\Delta_d(s) = s^4 + 10s^3 + 35s^2 + 50s + 24$$

Then, since $\Delta_o(s).s^r = s^4 + 6s^3 + 7s^2 + 2s$

we obtain $\underline{\delta}_1 = (4 \quad 28 \quad 48 \quad 24)$

and $$X_1 = \left(\begin{array}{c|cc|cc} 1 & 1 & 0 & 0 & 0 \\ 6 & 6 & 1 & 1 & 0 \\ 7 & 4 & 6 & 2 & 1 \\ 2 & 0 & 4 & 0 & 2 \end{array}\right)$$

A g_1-inverse of X_1 is

$$X_1^{g_1} = \begin{pmatrix} 1 & 0 & 0 & 0 \\ 0 & 0 & 0 & 0 \\ 3 & -1 & \frac{1}{2} & -\frac{1}{4} \\ -9 & 2 & -\frac{1}{2} & \frac{1}{4} \\ -7 & 2 & -1 & 1 \end{pmatrix}$$

and this results in $X_1 X_1^{g_1}\,\underline{\delta}_1 = \underline{\delta}_1$.

Thus, the required coefficient vector $\underline{p}_1$ is determined as

$$\underline{p}_1 = X^{g_1}\underline{\delta}_1 = \begin{pmatrix} \gamma_1 \\ \hline \theta_{10} \\ \theta_{11} \\ \hline \theta_{20} \\ \theta_{21} \end{pmatrix} = \begin{pmatrix} 4 \\ \hline 0 \\ 2 \\ \hline 2 \\ 4 \end{pmatrix}$$

i.e. $$\underline{m}^t(s) = \frac{1}{s+4}\,[2 \quad 2(s+2)]$$

This latter result illustrates the case where $X_1 X_1^{g_1} = I_4$.

S.10.5 Since $m+\ell-1 = 3 = n$, a full rank constant output-feedback compensator F can be designed in two stages such that the resulting closed-loop system has poles $\{\gamma_i\} = \{-1, -2, -5\}$.

(1) For $\underline{f}_c^{(1)} = (0 \quad 1)^t$, the pair $[A, B\underline{f}_c^{(1)}]$ is completely controllable, and

$$\underline{g}^{(1)}(s) = \frac{1}{s^3}\begin{pmatrix} 1 \\ s \end{pmatrix}$$

Here, $B\underline{f}_c^{(1)}$ influences all 3 modes of the system; i.e. $n_1 = 3$; however, there are only 2 linearly independent outputs; i.e. $\ell_1 = 2$; therefore, $q_1 = \min\{n_1, \ell_1\} = 2$. So, we determine $\underline{f}_o^{(1)}$ to place two poles at s = -1, -2, say, which results in the equations

$$(s^2+3s+2)(s+e_1) = s^3 + \{f_1(s^2+s+1) + f_2(s^2+s)\}$$

Solving for $\underline{f}_o^{(1)}$ and e_1, we obtain

$$\begin{pmatrix} 0 & 0 & -1 \\ 0 & 1 & -3 \\ 1 & 0 & -2 \end{pmatrix}\begin{pmatrix} f_1 \\ f_2 \\ \hline e_1 \end{pmatrix} = \begin{pmatrix} 3 \\ 2 \\ 0 \end{pmatrix}$$

which yields $e_1 = -3$,

and $$\underline{f}_o^{(1)} = (-6 \quad -7)$$

i.e. $$F_1 = \underline{f}_c^{(1)} \underline{f}_o^{(1)} = \begin{pmatrix} 0 & 0 \\ -6 & -7 \end{pmatrix}$$

Now, the resulting closed-loop system is

$$H^{(1)}(s) = \begin{pmatrix} s^2+s-6 & 1 \\ s^2+s+6 & s \end{pmatrix} \frac{1}{(s+1)(s+2)(s-3)}$$

(2) To hold one of these poles, say $s = -1$, at this value during the second stage of the design, we let $H^{(1)}(s) = \Gamma^{(1)}(s)/d_1(s)$ and solve

$$\underline{f}_o^{(2)} \Gamma^{(1)}(-1) = \underline{o}^t$$

i.e. $$\underline{f}_o^{(2)} \begin{pmatrix} 1 \\ -1 \end{pmatrix} = 0$$

$\therefore$ $$\underline{f}_o^{(2)} = [1 \quad 1]$$

Thus, $$N^{(2)}(s) = \underline{f}_o^{(2)} \Gamma^{(1)}(s)$$

$$= [2s(s+1) \quad (s+1)]$$

Now, to place the two remaining poles at $s = -1, -5$, we determine $\underline{f}_c^{(2)}$ by solving the equations

$$\begin{pmatrix} 2 & 0 \\ 0 & 1 \end{pmatrix} \begin{pmatrix} f_1 \\ f_2 \end{pmatrix} = \begin{pmatrix} 8 \\ 16 \end{pmatrix}$$

i.e. $$\underline{f}_c^{(2)} = \begin{pmatrix} 4 \\ 16 \end{pmatrix}$$

Thus, $$F_2 = \underline{f}_c^{(2)} \underline{f}_o^{(2)} = \begin{pmatrix} 4 & 4 \\ 16 & 16 \end{pmatrix}$$

Now ,

$$F = F_1 + F_2 = \begin{pmatrix} 4 & 4 \\ 10 & 9 \end{pmatrix}$$

which has rank 2 and generates the desired closed-loop system poles $\{\gamma_i\} = \{-1, -3, -5\}$.

Chapter 11: Solutions

S.11.1 From the example in Section 11.2, the fundamental term in the output of the relay is

$$\frac{4}{\pi}\sin(\omega t)$$

∴ the describing function is

$$N(U) = \frac{4}{\pi U}$$

∴ the graph of $-1/N(U)$ lies along the negative real axis, tending to $-\infty$ as $U \to \infty$.

The Nyquist plot of $g(s)$ is obtained from

$$g(j\omega) = \frac{1}{j\omega(4+5j\omega-\omega^2)}$$

which crosses the real axis at $-1/20$ when $\omega = 2$.

∴ the graphs intersect at the point corresponding to $N(U) = 20$, i.e. $U = 1/5\pi$.
Also, as U increases, the graph of $-1/N(U)$ passes out of the region encircled by the Nyquist plot.

∴ a stable limit cycle is predicted, with amplitude

$$U = 1/5\pi$$

and period

$$\frac{2\pi}{\omega} = \pi$$

S.11.2 The time-derivative of the Lyapunov function along the trajectories of the system is

$$\dot{V} = 2(x_1\dot{x}_1 + x_2\dot{x}_2)$$

$$= 2(x_1-1)x_2^2$$

$\therefore$ $\dot{V} \leq 0$ when $x_1 \leq 1$. Also, in the interior of this region, $\dot{V}$ can only vanish when $x_2 = 0$ $\therefore$ $\dot{V}$ does not vanish identically along any trajectory.

$\therefore$ any contour of V lying in the region $x_1 \leq 1$ encloses a domain of attraction.

$\therefore$ the largest domain of attraction obtainable with this Lyapunov function is

$$x_1^2 + x_2^2 < 1$$

Chapter 14: Solutions

S.14.1 By examining the form of the functional $J(u)$ in eqn. (14.13) we see

$$W = 0, \quad G = I, \quad R = 1$$

Also, since $b(x) = \sqrt{2/\ell}\sin \pi x/\ell$, $b_1 = 1$ and $b_i = 0$ $i > 1$

Substituting in eqn. (14.15) yields

$$\dot{p}_{mn} - \left(\pi^2\alpha/\ell\right)(n^2+m^2)p_{mn} = p_{1m}\, p_{n1}$$

$$p_{mn}(t_1) = 1 \qquad m = n$$

$$= 0 \qquad m \neq n$$

Clearly $p_{mn} = 0$ $m = n$ satisfies these equations. Then if $n > 1$

$$\dot{p}_{nn} - \left(2\pi^2\alpha n^2/\ell\right)p_{nn} = 0$$

$$p_{nn}(t_1) = 1$$

So

$$p_{nn}(t) = \exp(-2\pi^2\alpha n^2/\ell)\;(t_1 - t)$$

For $n = 1$

$$\dot{p}_{11} - \left(2\pi^2\alpha/\ell\right)p_{11} = p_{11}^2$$

So

$$\int \frac{dp_{11}}{p_{11}(p_{11}+2\pi^2\alpha/\ell)} = \int dt$$

Thus $$\left(\ell/2\pi^2\alpha\right)\log\left(\frac{p_{11}}{p_{11}+2\pi^2\alpha/\ell}\right) = t + \text{constant}$$

or

$$\frac{p_{11}}{p_{11}+2\pi^2\alpha/\ell} = Ae^{\frac{2\pi^2\alpha t}{\ell}}$$

But $p_{11}(t_1) = 1$, so

$$p_{11}(t) = \frac{\frac{2\pi^2\alpha}{\ell}\, e^{-\frac{2\pi^2\alpha}{\ell}(t_1-t)}}{1 + \frac{2\pi^2\alpha}{\ell} - e^{-\frac{2\pi^2\alpha}{\ell}(t_1-t)}}$$

The optimal control is

$$u^*(t) = -p_{11}(t) \langle \phi_1(\cdot), z(\cdot,t) \rangle$$

$$= -p_{11}(t) \int_0^\ell \sqrt{2/\ell}\, \sin \pi x/\ell \; z(x,t)dx$$

Since $A^* = A$, using equations (14.17) and setting

$$V = 1, \quad c = b, \quad P = I$$

we obtain the following dual filtering problem,

$$\frac{\partial T}{\partial t}(x,t) = \alpha \frac{\partial^2 T}{\partial x^2}(x,t)$$

$$y(t) = \int_0^t \int_0^\ell \sqrt{2/\ell}\, \sin \pi x/\ell \; T(x,s)dxds + v(t)$$

$$T(x,0) = T_o(x)$$

where the covariance of $v(t)$ and $T_o(s)$ is 1 and I respectively and the Kalman-Bucy filter is

$$\hat{T}(x,t) = \int_0^t \sqrt{2/\ell}\, e^{-\frac{\alpha\pi^2}{\ell}(t-s)} \; p_{11}(t_1-s) \sin \pi x/\ell \; d\rho(s)$$

where

$$\rho(t) = y(t) - \int_0^t \int_0^\ell \sqrt{2/\ell}\, \sin \pi x/\ell \; \hat{z}(x,s)dxds$$

S.14.2 Taking Laplace transforms we have

$$s^2\ \bar{w}(x,s) = \frac{d^2\bar{w}}{dx^2}(x,s) + \hat{b}(x)\bar{u}(s) \qquad (14.18)$$

$$\bar{y}(s) = \int_0^1 \hat{c}(x)\bar{w}(x,s)dx$$

$$\bar{w}(0,s) = \bar{w}(1,s) = 0 \qquad (14.19)$$

Let us consider the function

$$w(x,s) = \tfrac{1}{2}(e^{sx}(\alpha + \int_0^x e^{-s\rho}\ \hat{b}(\rho)d\rho)$$
$$+ e^{-sx}(\beta - \int_x^1 e^{s\rho}\ \hat{b}(\rho)d\rho)\ \bar{u}(s)$$

Then

$$\frac{d^2\bar{w}}{dx^2}(x,s) = s^2\bar{w}(x,s) + \hat{b}(x)\bar{u}(s)$$

So this function will satisfy (14.18), (14.19) if $\bar{w}(0,s) = \bar{w}(1,s) = 0$ and this will be the case if

$$(\alpha+\beta - \int_0^1 e^{s\rho}\ \hat{b}(\rho)d\rho) = 0$$

$$(e^s(\alpha + \int_0^1 e^{-s\rho}\ \hat{b}(\rho)d\rho) + e^{-s}\ \beta) = 0$$

so that

$$= \frac{1}{1-e^{2s}} \int_0^1 (e^{s\rho} + e^{s(2-\rho)})\hat{b}(\rho)d\rho$$

$$= \frac{1}{1-e^{-2s}} \int_0^1 (e^{s\rho} + e^{-s\rho})\hat{b}(\rho)d\rho$$

This yields the transfer function

$$\frac{1}{2}\int_0^1 \hat{c}(x)(e^{sx}(\alpha + \int_0^x e^{-s\rho}\ \hat{b}(\rho)d\rho)$$
$$+ e^{-sx}(\beta - \int_x^1 e^{s\rho}\ \hat{b}(\rho)d\rho)\ dx$$

In order to find the semigroup we have to solve the

inhomogeneous equation, but first we write the equation in abstract form

$$\dot{z} = Az + bu$$

$$y = \langle c, z\rangle$$

$$z_1(x,t) = w(x,t) \;;\quad z_2(x,t) = \frac{\partial w}{\partial t}(x,t)$$

$$b(x) = \begin{pmatrix} 0 \\ \hat{b}(x) \end{pmatrix}$$

$$c(x) = (\hat{c}(x), 0)$$

$$Az = \begin{pmatrix} 0 & I \\ \frac{d^2}{dx^2} & 0 \end{pmatrix}\begin{pmatrix} z_1 \\ z_2 \end{pmatrix}$$

and

$$D(A) = \{z \in H_0^1(0,1),\ z_2 \in H^2(0,1) \cap H_0^1(0,1)\}$$

In order to solve the inhomogeneous equation we will assume the solution separates so that

$$w(x,t) = T(t)X(x)$$

then
$$\frac{1}{T(t)}\frac{d^2T}{dt^2}(t) = \frac{1}{X(x)}\frac{d^2X(x)}{dx^2} = \lambda^2$$

So that

$$T(t) = A \sin \lambda t + B \cos \lambda t$$

$$X(x) = C \sin \lambda x + D \cos \lambda x$$

If the boundary conditions are to be satisfied $D = 0$ and $\sin \lambda = 0$. So $\lambda = n\pi$, and the complete solution has the form

$$w(x,t) = \sum_{n=1}^{\infty} (A_n \sin n\pi t + B_n \cos n\pi t) \sin n\pi x$$

If $w(x,0) = w_o(x)$ and $\frac{\partial w}{\partial t}(x,0) = w_1(x)$, we find

$$B_n = \sqrt{2}\ \langle w_o, \phi_n\rangle \qquad \phi_n(x) = \sqrt{2} \sin n\pi x$$

$$n\pi A_n = \sqrt{2}\ \langle w_1, \phi_n\rangle$$

So

$$w(x,t) = \sum_{n=1}^{\infty} (\langle w_o,\phi_n\rangle \cos n\pi t + \frac{1}{n\pi} \langle w_1,\phi_n\rangle \sin n\pi t)\phi_n(x)$$

But the full state is $z(x,t) = \begin{pmatrix} w(x,t) \\ \frac{\partial w}{\partial t}(x,t) \end{pmatrix}$ so the semi-group is given by

$$T_t z_o = T_t \begin{pmatrix} w_o \\ w_1 \end{pmatrix} = \begin{pmatrix} \sum_{n=1}^{\infty} (\langle w_o,\phi_n\rangle \cos n\pi t + \frac{1}{n\pi}\langle w_1,\phi_n\rangle \sin n\pi t)\phi_n \\ \sum_{n=1}^{\infty} (-n\pi \langle w_o,\phi_n\rangle \sin n\pi t + \langle w_1,\phi_n\rangle \cos n\pi t)\phi_n \end{pmatrix}$$

From (14.12) we see that the system is initially observable if

$$\sum (\langle w_o,\phi_n\rangle \cos n\pi t + \frac{1}{n\pi} \langle w_1,\phi_n\rangle \sin n\pi t)c_n = 0$$

for all $t \in [0,t_1]$ implies $w_o = w_1 = 0$. Clearly this will be the case if $\langle w_o,\phi_n\rangle = \langle w_1,\phi_n\rangle = 0$ for all $n = 1,2,\ldots$. If $t_1 = 2$ $\{\sin n\pi t, \cos n\pi t\}$ form a basis for $L^2(0,2)$ so in this case the system will be initially observable if $c_n \neq 0$ for all $n = 1,2,\ldots$.

From (14.11) we see that the system will be approximately controllable if

$$\sum_{n=1}^{\infty} (n\pi \langle w_o,\phi_n\rangle \sin n\pi t + \langle w_1,\phi_n\rangle \cos n\pi t)b_n = 0$$

for $t \in [0,t_1]$ implies $w_o = w_1 = 0$. Again this will be the case for $t_1 = 2$ if $b_n \neq 0$ for all $n = 1,2,\ldots$.

Index